FATAL FAILS

FATAL FAILS

Scientific Problems Inherent In The Theory Of Global Warming

Earl J. Seeley

Tampa, Florida

Fatal Fails:
Scientific Problems Inherent in
the Theory of Global Warming

Published by Gatekeeper Press
7853 Gunn Hwy, Suite 209
Tampa, FL 33626
www.GatekeeperPress.com

Library of Congress Control Number: 2023934354

ISBN (paperback): 9781662937248
eISBN: 9781662937255

Why This Book?

Why Now?

The control of the earth's climate, including the temperature of the earth's surface and near earth atmosphere, is a complex, intricate, and beautiful system of interactions between a number of forces. These controlling factors, their interactions, and the rate and extent of those interactions, are all variables which contribute to the control of the earth's climate, including its surface temperature.

I first became aware of the "Theory of Global Warming" in the late 1980's when I first found the document: (Hansen, J.E., and S. Lebedeff, 1987: Global trends of measured surface air temperature. *J. Geophys. Res.*, **92;**13345-13372).

As a fruit grower during my youth; A professional horticultural researcher who had worked on different methods of frost control in fruit crops, and in basic research in those areas as well as related subjects; I was acutely aware of the greenhouse effect, its causes, and importance in the control of earth's surface temperature. The temperature of plant tissues, under various climatic circumstances has been something that has intrigued me and been a subject of my research for over 50 years.

The small, but significant, increase in the earth's surface temperature, over the last 110 years has been attributed to slight increases in the concentration of carbon dioxide, methane, and other minor greenhouse gasses.

The danger to the world's societies, of our present course, the result of attributing all the warming of earth's surface to the greenhouse impact of these minor greenhouse gasses, is real and alarming.

If we utilize most of our available resources in limiting or decreasing the concentration of these minor greenhouse gasses, we will not have the resources required to deal with the real impacts of the major forces behind the warming of earth's surface, should they continue or increase.

In this work you will be exposed to the truth about the flaws and fails of the global warming/ climate change theory. You will read about, and see peer reviewed research that establishes the impacts of:

1. The three forces which have increased the intensity of the sunlight reaching and warming the earth's surface.
2. The role of the LaNina/ElNino cycle and the North Atlantic Oscillation in causing a major portion of the ice volume loss in, and warming of, the Arctic.
3. The role of increasing levels of heat from the earth's core in causing significant ice volume and extent losses, across the Arctic and Antarctic.
4. Changes in the amount of water vapor, liquid and ice in the atmosphere over long periods of time, and this forces contribution to changes in climatic patterns and earth's surface temperature.
5. You will also be exposed to the climatic impacts of the increasing aerosol concentration in the atmosphere.

The above listed causes of changes in the earth's surface temperature have all been mistakenly attributed to the greenhouse effect of the minor greenhouse gasses because of the basic Fatal Fails of the "Global Warming/Climate Change Theory". These Fatal Fails are identified and discussed in this work.

Photo: The Bald Eagle is a symbol of many things. This magnificent specimen was photographed 7 miles above Chief Joseph Dam, along the Columbia River, from a concealed position in a row of popular trees. After I took this photo, I carefully withdrew, moved approximately 150 yards to the left and advanced, crawling until I could once again see this bird. The photo I took from that concealed position, was nearly identical to this one; except, it shows a view of his other eye. He sees his environment as it actually is. He knew what and where I was in spite of my best efforts at concealment. That is our challenge, we need to see the world as it actually is, not as we are told it should be, or is, by someone who may have a stilted or myopic view of the earth and our environment.

Prologue:

There is no such thing as settled science! To assume or declare that there is, opens the door to abuse of scientific methods, interjection of personal biases, and the sorry state of affairs that currently grips much of the scientific community. Recently a member of a State Universities' Climate Sciences Department told me: "Every time there is a big weather event, or other problem there is a rush to the computers to be the first to find a connection to Global Warming". I see many examples where the above statement would be more accurate if the word "find" was replaced with the word "create".

During my formal, scientific educational experience, spanning the years of 1961 - 1997 (when the years spent as a Research and Graduate Facility member and adjunct Professor are included), I was often exposed to, studied, and taught the "1st and 2nd Laws of Thermodynamics". These two concepts: 1st - The Law of the Conservation of Matter - That matter cannot be created or destroyed, and 2nd - The Law of Entropy - That all matter tends toward a random distribution, had special status. They were "Laws," because we knew and understood them. Looking back it is abundantly clear that we gave them undeserved reverence, because we now understand them differently.

When a true scientific principle is involved, the degree of perfection in our understanding and interpretation of it, does not change the basic nature of the principle. In my view, there are some basic unalterable truths (none of which, we can presume to understand perfectly), which operate to control this earth and the universe in which we live. What we think, say, or teach does not change the nature and truth of these principles.

For example: No matter what one may think or believe about gravity it will function in fixed patterns. The photo on p. 5, is of a section of the trail to Hidden Canyon in Zion National Park in southern Utah (USA). I have hiked it many times, almost always in the late fall or very early spring, when there are no crowds. There is a chain affixed to the wall along the section shown in this photo. If you hike this trail use the chain and if you must pass someone going the opposite direction, use extreme caution. From time to time someone has made a mis-step on this trail and gone over the edge with predictable results.

Gravity accelerates the speed of the body in fixed ways, to the point that it is impossible to survive the sudden stop on the rocks 500 feet below. It happens every time. It has happened a significant number of times. A free fall in this circumstance has only one end-point (a basic, unalterable truth).

We must constantly be re-examining what we think we know, while holding onto the safety chain of objectivity. To cease to "look" and examine what we think we know for bias is equivalent to moving along this trail with our eyes closed and not holding onto the chain.

We must always guard against: "The Normalization of Deviance and Groupthink" (See Appendix #1: p. 260.)

For more on these topics visit: <fatalflaws.net>
You may submit questions or comments by email to: <earljseeley@gmail.com> Some I will answer on the website <fatalflaws.net> others by return email.

FATAL FAILS: Scientific Problems Inherent in the Theory of Global Warming

Table of Contents

FATAL FAILS: Scientific Problems Inherent in the Theory of Global Warming

Book Abstract: The assumption that, an increase in the CO_2 content of the atmosphere, equal to 0.121 hundredths of 1% of the atmosphere's dry weight, is solely responsible for all the warming of the earth that has occurred over the last 140 years is preposterous. There are other environmental forces which are much stronger, prevalent, and active in regulating and modulating the earth's surface temperature than 400 ppm of atmospheric carbon dioxide (or the 121 ppm increase over the period from 1880 to 2020). These forces are: First, the sun; Second, the earth's waters; Third, heat from the earth's core; and Fourth, airborne aerosols.

All of these forces vary over time and have contributed to the surface and near earth atmospheric temperature increases that have occurred over the last 110 years. They, in the aggregate, dwarf the impact of the 121 ppm increase in the CO_2 concentration of the atmosphere, part of which is attributable to human activities.

Carbon dioxide is an extremely important substance in the support of life on this planet. It plays several very important roles, but its contribution to the greenhouse effect of the atmosphere is a "bit part," not a "leading role."

In the first paragraph above I stated: "The assumption that, an increase in the CO_2 content of the atmosphere equal to 0.121 hundredths of 1% of the atmosphere's dry weight, is responsible for all the warming of the earth that has occurred over the last 110 years is preposterous". The central reason that this statement is true, is that the assumption involves a single dimension interaction, consistent with the single dimension models used in climate science to define the relationship between the concentration of CO_2 and the earth's surface temperature.

As also stated above, there are four other major factors that are driving and interacting processes involved in controlling the earth's surface temperature. A single dimension model can't describe or project outcomes, which involve the interactions of five different factors, variables or forces.

Let me describe as precisely as possible how a single dimensional interaction works and contrast that with a multi-dimensional interaction.

There are Five main variables involved in the control and modulation of earth's surface temperature. These variables are, in order of importance: 1. Variations in the intensity of the Sun's irradiance reaching the earth's surface; 2. Water, and variations in the amount of water in the atmosphere; 3. Heat from earth's core; 4. Aerosols suspended in the atmosphere; and 5. The minor greenhouse gases; CO_2 being the most important of the group.

Viewed through a single-dimensional model between Temperature and CO_2 concentration, with the basic assumption that; "Any heating will be the result of CO_2 forcing," let us examine the contents of a Mason Jar:

We begin with a 1 quart Mason Jar, containing 1 pint (473 g) of water, at a temperature of 20^0C.

A. We add a net of 473 calories of heat as radiant energy from CO_2.

B. The result is that the water now has a temperature of 21^0C.

C. The conclusion: The CO_2 has caused a 1^0C rise in the water temperature.

D. We now add a net of 473 calories of heat as sunlight energy.

E. The result is that the water now has a temperature of 22^0C.

F. The conclusion: The CO_2 has caused a 2^0C rise in the water temperature.

Does the above make sense? The answer is no; but, remember the controlling assumption: "all heating will be the result of CO_2 forcing of the contents of the Mason Jar"; Hence the conclusion, no matter how ridiculous; "The heating is all the result of CO_2 forcing". If the heat came from water, magma, or the greenhouse impact of aerosols the conclusion would be the same, forced by the underlying assumption; "The heating is all the result of CO_2 forcing."

<u>The Global Warming/Climate Change theory, because of this attribute of single dimensional models, lumps the warming contributions of all other contributing forces together under the carbon dioxide forcing moniker.</u>

Viewed through a multi-dimensional model between temperature and CO_2, with the basic assumption that any heating of the contents of a Mason Jar may be the result of energy from; 1.The sun and/or; 2. Water, and/or; 3. Heat from earth's core; and/or 4. Aerosols, and/or; 5. Carbon dioxide - - -

We begin with a 1 quart Mason Jar containing 1 pint (473 g) of water at a temperature of 20^0C.

A. We add a net of 473 calories of heat as radiant energy from CO_2.

B. The result is that the water now has a temperature of 21^0C.

C. The conclusion: The CO_2 has caused a 1^0C rise in the water temperature.

D. We now add a net of 473 calories of heat from sunlight energy.

E. The result is that the water now has a temperature of 22^0C.

F. The conclusion: CO_2 has caused a 1^0C rise in the water temperature, and the Sun has caused a 1^0C rise in the water temperature.

If the heat added in A, or D above resulted from sitting the jar in warm water, or on cooling magma, or from the greenhouse effect caused by aerosols, the added heat would be credited to the proper source. That does not happen with a single dimension, myopic, modeling system. It is all lumped together as CO_2 forcing.

Authors note: I realize that the above seems unbelievable, but it is true. Rhetorical question: Have you ever seen an academic or government source which breaks down (by percentage or other proportioning system) a series of different forces that contribute to the earth's warming. This is the greatest Fatal Flaw that leads to the Fatal Fails of the Global Warming/Climate Change Theory.

When it is broken down it is easily understood, and the fatal fail forced by the ridiculous underlying assumption is exposed.

As you read through this work you will come to realize how important the above concept is to understanding how nature works.

In addition to the above described preposterous assumption, there are six fatal fails at the heart of the global warming theory. Each is discussed in this book, and supporting evidence presented for the person(s) who desires to know the truth.

These fatal fails occurred, when in forming the “Theory of Global Warming," the contributions of four of the most important environmental factors in the control of the earth’s surface temperature were assumed to have had a net effect on global warming, that is equal to zero. The remaining two fatal fails concern CO_2’s roles and fate in nature.

The six Fatal Fails are:

Fatal Fail 1: The assumption that, the impact of variations in the suns radiant energy that reaches the earth’s surface has had no net effect on the warming of earth’s surface over the last 110 years. (The global warming period is most often represented as 140 years {1880 to the present}, but the first 30 years of that period saw earth’s surface temperature anomaly decrease from -0.16^{0}C (1880) to -0.44^{0}C (1910), while the CO_2 concentration rose by 11 ppm {a negative correlation period between CO_2 concentration and earth’s surface temperature}).

Fatal Fail 2: The assumption that, the role of water in controlling and modulating the earth’s surface temperature and the variations in the quantity of water in the atmosphere, has had no net impact on the earth’s warming over the last 110 years.

Fatal Fail 3: The assumption that, the effect of heat from the earth’s core in the forms of volcanic activity, magma flows, thermal vents, heat moving through the earth’s crust and interactions between that heat and the earth’s waters has had no net impact on the earth’s warming over the last 110 years.

Fatal Fail 4: The assumption that, the role of increasing airborne particulate and aerosols in the greenhouse effect has had no net impact on the earth's warming over the last 110 years.

Fatal Fail 5: The role of increasing CO_2's greenhouse impact has been vastly over-estimated. The increase in the level of CO_2 since 1880 equals 121 ppm, or 1/83 of 1% of the earth's atmosphere. (This fatal fail of the theory results from the use of single dimension models, which lump all heating of earth's surface under the CO_2 moniker.)

Fatal Fail #6: The dynamic usage and sequestration of CO_2 by the photosynthetic process, and 3 physical-chemical processes that occur 24/7/365, over much of the earth's surface, sequesters most of the CO_2 added to earth's atmosphere each year by human activities. Of the 47* billion tons (2020 figure) of CO_2 released into the atmosphere annually by human activities, all but 14.85* billion tons are removed and sequestered. (*These figures are adjusted upward, to account for the significant increase in the amount of CO_2 released by human activities during the 2018 to 2021 period by increased wildfire destruction. See Chapter 9)

Example #1 of how these flaws have led to major fatal fails of the theory; On November 5, 2018, a study, was published by staff and collaborators of the Lawrence Livermore National Laboratory. It concluded, that over the last 3.8 decades (38 years), between 40% and 50% of all the Arctic ice melt that has occurred, has been the result of natural variation related to El Nino/La Nino events, not to the four one hundredths of 1% (that is 1/25 of 1%) of the earth's atmosphere that is CO_2. This ice loss is not the result of human activities, it is natural variation.

Example #2 of how these flaws have led to major fatal fails of the Theory: From: NSF site: <https://

www.nsf.gov/news/news_summ.jsp?cntn_id=295861> June 27, 2018. "Tracing a chemical signature of helium in seawater, an international team of scientists funded by the National Science Foundation (NSF) and the United Kingdom's (UK) Natural Environment Research Council (NERC) has discovered a previously unknown volcanic hotspot beneath the massive West Antarctic Ice Sheet (WAIS). Researchers say, "The newly discovered heat source could contribute in ways yet unknown to the potential collapse of the ice sheet."

Before we complete this work you will see numerous other examples of sound scientific research, peer reviewed and published, that will illustrate that most of the warming of the earth's surface that has occurred since 1880 has been caused by things, other than the greenhouse effect of the minor greenhouse gasses.

So why is the Global Warming/Climate Change Theory so widely accepted? See Appendix 1: "The Normalization of Deviance and Groupthink" p. 260. Also visit: <fatalflaws.net>

Introduction

This book addresses the false science behind what has been represented to be the "settled science" of global warming. We will begin by addressing the six Fatal Fails of the Global Warming Theory described on pages 11 and 12 above. In a series of six chapters we will discuss:

First: How changes in the amount of solar energy that has reached the earth's surface between 1880 and the present, has contributed to the warming of the earth.

Second: How the miraculous physical properties and vast distribution of water, over much of the earth's surface, and changes in the quantity of water in the atmosphere, has caused significant changes in climatic patterns over the earth's surface during the last 110 years. It is key to understand that these changes are not caused by CO_2 (and the other minor greenhouse gasses). They are what is termed "natural variation".

Third: How heat from the earth's core, rising to the surface in the forms of volcanic activity, magma flows, thermal vents, and along tectonic plate margins has contributed to the warming of the earth's surface.

Fourth: How the significant increase in the aerosol content of the earth's atmosphere has resulted in a slight cooling of the earth, which is included in the energy budgets that support the global warming theory. These particles have also contributed to the warming of earth's surface because of their role in the greenhouse effect, something ignored by the global warming advocates. Ninety percent of these particles have natural origins (according to NASA).

Fifth: We will closely examine how and why the role of carbon dioxide in the warming of the earth's surface over the last 140 years has been grossly inflated.

Sixth: We will closely examine the dynamic flow of atmospheric carbon dioxide through the biosphere and how great quantities of this gas are removed from the atmosphere and sequestered for long and very long periods of time.

Following the six chapters described above we will discuss how a significant portion of the carbon dioxide accumulated in the atmosphere is not "human caused". We will also discuss how deforestation, and changes in wildfire prevention, control and remediation have resulted in significant increases in atmospheric carbon dioxide.

We will end this work with a thorough examination of the global warming numbers. This will include, how the actual data is not used in the determination of the global mean temperature. The actual data is processed through an increasingly complex series of steps that yield what the global warming advocates refer to as "the unadjusted data" (or the data they use).

The "processing of data" referred to in the previous paragraph is unprecedented in "climate science" or any science. Without the blatant fudging of the numbers, the magnitude of the warming of the earth's surface would be approximately 50% of the current claims.

Before we move forward there is one more "housekeeping item" we need to illustrate and clarify. That is: <u>How much has the atmospheric carbon dioxide increased during the "global warming" period?</u>

Most people don't comprehend the CO_2 concentration numbers involved in the global warming theory. The pre-industrial period level of 290 ppm of CO_2 represents 0.290 one hundredths of 1% of the earth's atmosphere's dry weight. The current level of 411 ppm represents 0.411 one hundredths of 1% of the earth's atmosphere (0.411 one hundredths of 1% is equal to 1/25 of 1%). The increase of 121 ppm represents only 0.121 one hundredths of 1% (or 1/83 of 1%) of the dry weight of earth's atmosphere.

To put these figures into perspective, assume that we have 1,000 pounds of Sugar, which weighs 50 pounds per cubic foot. So the sugar cube would be two feet by two feet by five feet high, for a total volume of 20 cubic feet.

This cube weighs 16,000 ounces and contains 96,000 teaspoons of sugar (6 tsp./oz). One-hundredth of 1% of this cube would be equal to 9.6 teaspoons. Multiply 9.6 teaspoons of sugar by 0.121 and you have 1.152 teaspoons out of the 96,000 teaspoons of sugar. This is the ratio of the increase in the CO_2 content to the other gasses in the atmosphere. That is 1.152 to 95,998.848. I am not a coffee drinker, but I don't think that is enough to make a single cup very sweet.

The total CO_2 content of the atmosphere; 0.04%, would be equal to 3.84 teaspoons. Neither of these figures are sufficient to account for the claimed effects on the earth's surface temperature.

These figures represent the percentage of the earth's atmosphere, if it existed in a waterless (dry) state. In this planet's circumstance, air does not naturally exist with no water content (humidity) in the lower 40,000 feet.

The total water vapor molecules in the atmosphere is equal to approximately 24.4 times that of CO_2. This does not consider the liquid and solid state water which makes up the clouds, fog, snowflakes and raindrops. We will discuss, in Chapter 3, why this means that CO_2, ozone, methane, nitric oxide, and other non aqueous gasses, in total, are minor contributors to the greenhouse effect.

Photo (p.17): My home is at an elevation of 6550 feet. A few years ago, a huge dust storm that originated from the Gobi desert in China, moved huge amounts of dust (particulate) into the atmosphere. A few days later I observed and photographed this sunrise. The interaction between the suns rays and the particulate (aerosols) in the atmosphere produced the most vivid sunrise I have observed in my 79 years.

Chapter 1: The Sun's Role in Determining the Earth's Surface Temperature

Synopsis: *The sunlight that reaches the earth is, without question, the most important factor in determining the earth's surface temperature. The distance of the earth from the sun, and the tilt of the earth's axis, which determines the seasons, are nearly constant. Changes in the temperature of the sun will cause variations in the suns impact on the earth. So the question is; Does the temperature of the sun change? The answer to this question is; yes, it does.*

Forces within the sun result in cyclic increases and decreases in the number of sunspots and solar flares. Changing the number of sunspots and solar flares, changes the temperature of the solar surface and the amount of energy emitted by the sun that reaches the earth.

There was a gradual increase in sunspots and solar flares between 1900 and 1940. Then there was a huge increase in the early 1950s and solar activity remained historically high up to 2018. There has also been a gradual but significant increase in the temperature of the sun during the "quiet sun" stage of the solar cycles.

The photo below-left, shows the sun during the solar minimum of 2010 (quiet stage) when the sunspot numbers were only 1 to 3 per month. The photo below-right, shows the sun during October 2012 when cycle 24 first reached 100 sunspots per month. Approximately 4 years later this cycle peaked at 120 sunspots per month. It then fell to near zero for the entire period from 2018 to Feb. 2021. (NASA SDO: 2010 and 2012 comparison)

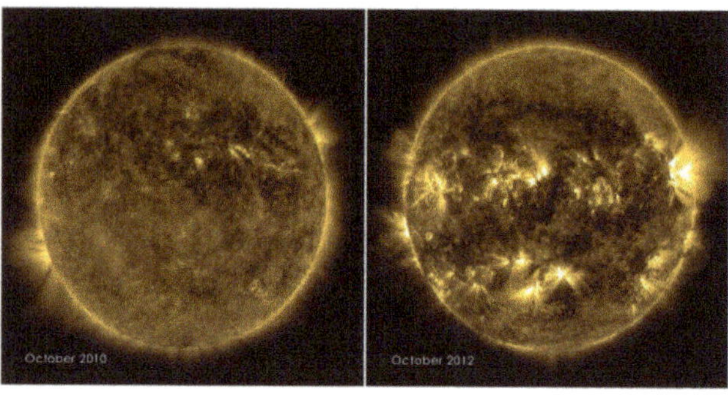

Early in the space-age (circa 1976), a satellite orbiting 200 miles above the earth measured the intensity of the solar irradiance at the "outer limits" of the atmosphere. By todays standards those measurements were fairly crude but they were quite accurate. Since that time we have been able to monitor the temperature of the sun, and the solar energy that reaches earth's outer atmosphere.

In this chapter we will examine three mechanisms or processes which have increased the amount of solar irradiance which has reached *and been absorbed by the earth's surface. These processes have nothing to do with CO_2, yet their contribution to the warming of the earth's surface has been credited to CO_2 temperature forcing of the earth's surface.*

We will examine why and how the erroneous crediting of these warming influences has been attributed to the increased CO_2 content of earth's atmosphere. This is an extremely important fatal fail of the global warming theory.

Photo: One of the many things the sun adds to this earth is beauty. This sunset behind 12,000' Mt. Nebo in central Utah, USA is an example of the enriching influences we experience as part of life on this wonderful orb.

<u>*Chapter 1:*</u> No one can deny that the sunlight that reaches earth, plays the dominant role in determining the earth's surface temperature. The distance of earth from the sun, the tilt of the earth on its axis and the seasonal variation that tilt causes, are the primary factors that determine how warm the earth is. If the distance between earth and the sun was much less, earth's temperature would be too hot. If the distance

from the sun was much more the earth would be too cold.

If the earth was not tilted on its axis, much of the northern and southern tropical zones would be too hot. But for the tilt of the earth, the polar regions and northern portion of the northern hemisphere temperate zone would be much colder (as would the southern portions of the temperate zone in the southern hemisphere).

The role of distance from the sun, in determining the temperature of a planet, is illustrated by the fact that Mercury (closest planet to the sun) has a temperature of approximately 800^0F on the side facing the sun, and a temperature of -290^0F on the "night side". The Earth's average temperature is approximately 57^0F and Uranus, the 7th planet from the sun has an average temperature of -318^0F. This illustrates the relationship between distance from the sun and planet temperature. Yes, we live on a "Goldie Locks Planet," it is just right.

A little history is in order at this point. The Chart on page 21 from Encyclopedia Britannica – on line, shows the solar cycles from 1610 to 2004, a period of 394 years. The red vertical lines show the yearly number of sun spots.

After over 44 years of monitoring the intensity of the sun's rays at the outer limits of earth's atmosphere with satellite instrumentation, we know that when sun spot numbers are low, the amount of solar energy reaching the earth's atmosphere is lower than the long-term average.

We also know that when sunspot numbers are high, more energy reaches the outer limits of the atmosphere approximately 200 miles up, with X-rays and Ultraviolet energy increasing significantly. Of the solar energy which reaches the outer atmosphere, 24% is ultimately absorbed by the earth's surface (NASA and UN IPCC's figure).

It is significant that much of the increase in the solar irradiance from sunspots, is in the high energy region of ultraviolet light. These photons (energy packets) each carry two to four times the energy, per photon, that the red and infrared areas of the spectrum contain.

Examining the chart below, one sees that between the time the record began to approximately 1950, only four of the solar cycles exceeded 100 sunspots in a single month. The 1620, 1840, 1920 and 1940 cycles exceeded 100 but no cycle in this period averaged 130 sunspots for a single month. During this time, there were two extended periods when sunspot numbers were very low, the Maunder Minimum and the Dalton Minimum.

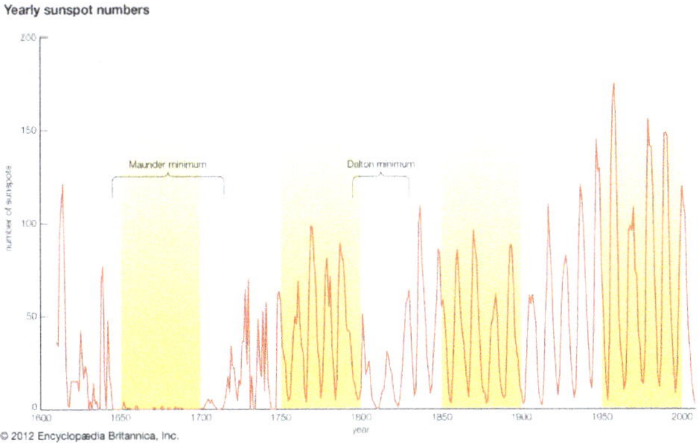

https://www.britannica.com/science/Dalton-minimum

The Maunder minimum spanned the coldest portion of the Little Ice Age in the Northern Hemisphere lasting from approximately 1640 to 1720. During this period the Vikings were forced to abandon their settlements on Greenland because there was no access by sea.

For a time, Iceland was not accessible by boat because the arctic ice sheet extended south of it year around. Glaciers became a problem for Norwegian Farmers and the Thames River in England regularly froze over.

The Dalton Minimum, lasting from 1790 to 1830, was a period of low solar activity (though not as low as the Maunder Minimum and for a shorter time). Both of these periods, of low solar activity, coincided with periods of significantly lowered global temperatures.

Some argue that the lower global temperatures of the Dalton minimum were intensified if not caused by the eruption of Mt. Tamboura in Indonesia. The 1815 eruption of Tamboura was the most powerful in recorded history, with a Volcanic Explosivity Index (VEI) of 7.

The weakness in the argument that the eruption of Mt. Tamboura caused the global temperatures to decrease, not the low solar activity, lies in four facts.

First: The Dalton Minimum began 20 years before the eruption of Mt. Tamboura and significant global cooling had already begun.

Second: While 1816 was known as the year without summer in the northern hemisphere because of the Mt. Tamboura ash in the atmosphere, the Dalton Minimum and a significantly lowered global temperature lasted for an additional 22 years.

Third: On July 15, 1991, Mt. Pinatubo in the Philippines explosively erupted. This eruption was assigned a VEI of 6. Mt. Tamboura was assigned a VEI of 7, somewhat larger but both were very significant eruptions. Mt. Saint Helens (May 18,1980) was assigned a VEI of 5. These numbers are significant because, assuming that the extent of the annual minimum of arctic ice cover is a good indicator of temperature trends, it is noteworthy that; following the 1991 eruption of Mt. Pinatubo the extent of the arctic ice cover remained nearly constant until 1996. The ice cover did not expand. <http://nsidc.org/arcticseaicenews/charctic-interactive-sea-ice-graph/100/>

Fourth: Typical of current published statements about the relationship between earth's surface temperature and solar cycle activity is the statement of Erik Gregersen in Encyclopedia Britannia: "However, the physical mechanism that explains how changes in activity of the Sun affect Earth's climate is unknown, and such episodes, however suggestive, do not prove that lower sunspot numbers produce cooling."

This statement is flawed. We do know the mechanism that explains how changes in the activity of the Sun affects Earth's climate. We have the temperature correlations between the low sunspot numbers of the Maunder Minimum, plus the Dalton Minimum, and Solar Cycles 20 and 24 each of which resulted in global cooling.

Both cycles 20 and 24 have occurred during a time period when we are able to monitor the level of solar irradiance at the outer limits of the atmosphere with satellite systems including SABER. We also have the ability to monitor the solar radiation received by earth's surface and the temperature of the earth's surface with MODIS. These 3 different systems (operating on unprocessed data they gather and record) are in close agreement. They also agree with the NOAA NWS unedited data and substantiate the fact that solar flares and sunspots do warm the earth's surface.

At: <NASA earth observatory>; Solar Radiation and Climate Experiment (SORCE) January 21, 2003, we can find a lot of very good information about the sun. This source comes from a time period before NASA began to succumb to the biases that are dictated by the Global Warming/Climate Change doctrine. It is archived material that is no longer being updated (you may read that as: not changed to preserve climate change dogma).

Under the heading: Solar Variability you will find the following facts:

Sunspots do change the temperature of the sun. As sunspot numbers increase, the overall temperature of the sun increases. This can be seen in, and is explained by, the chart on p. 25 along with the following explanatory quote.

"The 11-year solar cycle is manifested by the appearance and disappearance of large numbers of sunspots on the Sun's surface. The image series below shows the Sun at a wavelength of 656.3 nm, where Hydrogen emission just above the Sun's visible surface reveals increased energy coming from faculae (area around the sunspot). One image was taken every year from 1980 to 1989. 1980 was near a solar maximum and the Sun was active, while 1986 was near the minimum, and the Sun's surface was almost featureless. (Image adapted from, The Sun: a pictorial Introduction by P. Charbonneau and O.R. White)"

In the chart shown on p, 25, you will find: "The change in the Sun's yearly average total irradiance during an 11-year cycle is on the order of 0.1 percent or 1.4 watts per square meter." This agrees with what you will read later in this chapter, about how the increase in the number of sunspots has caused a significant portion of the heating of the earth, over the last 110 years.

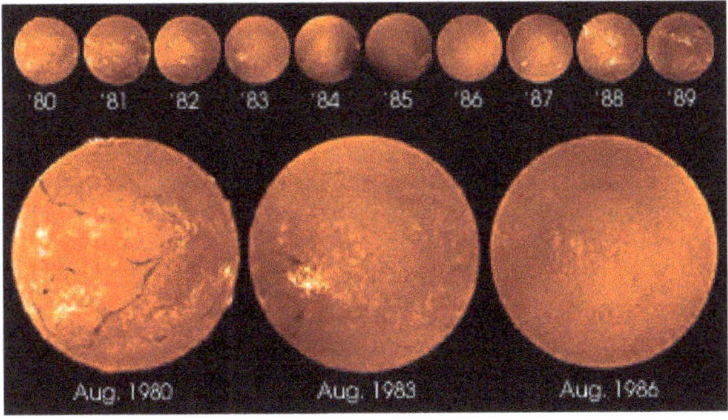

The above information source data, was received by the SORCE satellite. This is a different system than you will find referenced in other places in this work that are from SABRA and/or MODIS systems.

Authors Note: Both the SABRA and MODIS satellites use highly sophisticated spectroradiometer systems which can measure very narrow bands of the radiant spectra. This allows scientists to measure and identify many elements of solar irradiance and the earth's radiant emissions.

For example MODIS can measure the temperature of water surfaces by measuring specific wavelengths of the infrared spectrum. It can also measure the water temperatures at various depths using the intensity of certain microwave radiation bands.

SABRA can, by looking through the atmosphere toward the earth's surface, measure the specific wavelengths of radiation given off by CO_2 and nitric oxide. By measuring the intensity of specific wavelengths of energy given off by CO_2 and Nitric Oxide it can tell how much solar radiant energy is being absorbed by these gasses as the intensity of the radiant energy given off by the sun increases and decreases.

Forty-four years of direct monitoring, places the average total irradiance (sunlight energy) reaching the earth's outer atmosphere at 1,360 to 1,361.7 W/m^2 (Watts per square meter). The variation in the total has ranged by 7 to 9 Watts per square meter during the monitored 44 years (four solar cycles). That means that the sunlight reaching the earth's outer atmosphere varies by plus or minus 3.5 to 4.5 Watts per square meter during a strong solar cycle. These figures are measurements of the short term (daily) variations not the monthly or yearly averages. (You can verify this by checking the daily variation plot at: <NASA earth observatory>; Solar Radiation and Climate Experiment (SORCE) January 21, 2003.)

NASA and the IPCC both maintain that 24% of the sunlight that reaches the earth's outer atmosphere, reaches and is absorbed by the earth's surface. Simple math shows that the suns energy reaching the earth's surface varies by 0.84 to 1.08 Watts per square meter because of the suns temperature changes. This is a warming factor or force which global warming doctrine insists does not occur. It has been historically strong since the 1940s, the period when most of the earth's surface warming has occurred.

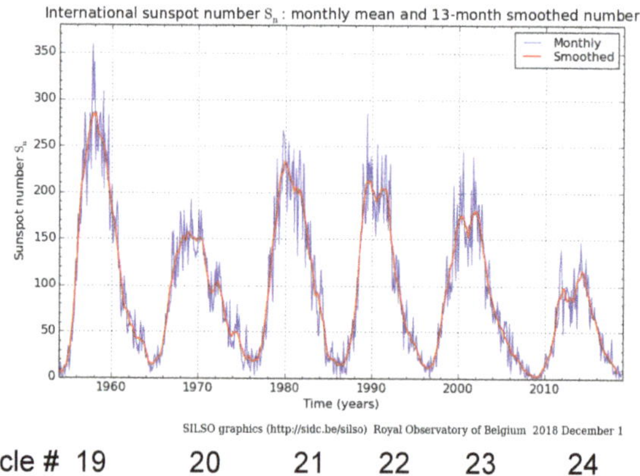

Cycle # 19 20 21 22 23 24

With the above in mind, examine the plot of sunspot numbers above and the Encyclopedia Britannica plot on p. 21 and note how active the sun has been since 1940. One cycle (19) exceeded 250 sunspots per month and persisted above that level for four years. Two cycles (21 & 22) exceeded 200 sunspots per month for three to four years, and one (cycle 20) reached 150 sunspots per month for extended periods. The immediate past cycle (cycle 24, with its extended, intense minimum) reached briefly 120 sunspots per month (the lowest maximum in 78 years).

To the current date (12/30/2022) the maximum sunspot number reached in cycle 25 has been 126 and the maximum monthly average is 113 sunspots and solar flares.

Caution: You will see figures from NASA and other sources that refer to declining solar activity over recent decades, they are correct in terms of, declining from the highest level in history (Cycle 19). They are not correct in terms of the longer term history involved in the Global Warming era (1880 to present). Also you will see NASA figures that refer to Cycle 24 reaching a maximum of 82 sunspots per month. This figure comes from LOWESS Smoothed data which is very misleading because of the smoothing effect. (One must compare apples to apples not apples to grapefruit.)

Looking at the vast difference in the pre-1940 and post-1950 solar activity (shown in the Encyclopedia Britannica Plot p. 21) you can see that even with a very small increase in the solar energy reaching the earth's surface one would expect a warming trend over the last 80+ years from this energy source. According to the Global Warming disciples the net effect of these higher levels of solar activity is equal to zero.

Quoting again the above referenced NASA source, under the heading "Solar Radiation and Global Warming: "Many researchers believe the steady rise in sunspots and faculae since the late seventeenth century may be responsible for as much as half of the 0.6 degrees of global warming over the last 110 years (IPCC, 2001)."

As mentioned above the increased sunshine reaching the earth's surface accounts for 50% of the energy that NASA and the UN's IPCC says is the excess incoming radiation versus the outgoing radiation (the heating factor). So the question is: Why isn't the earth warming faster than it is? The answer, is in other things the Global Warming Doctrine ignores or discounts.

For example, we will discuss in Chapter 3 how water accounts for at least 75% of all heat loss from the earth's surface. It does so in ways that avoid a large portion of the potential greenhouse impact of the minor greenhouse gases. This system is also temperature sensitive, as the earth's surface temperature decreases the system slows down; as the surface temperature increases, the system speeds up (known as a feedback, or buffering function).

Authors Note: The weak solar cycle 24 (a cooling factor) and El Nino and significant Volcanic activity (both warming factors) had been occurring at the same time (since 2004). This led to a weak correlation between decreasing sunspot numbers and cooling of the earth's surface temperature early in the cycle 24 to 25 minimum. This weakness was misused to challenge the correlation between sunspot activity and warming of the earth in publications such as Wikipedia and cited material there.

The current situation, which is a prolonged minimum period, followed by a weak beginning of cycle 25, is proving these publications wrong as we are now in the strongest cooling period since the space age began. I used the term "misused" in this paragraph because since 1610 when sunspot numbers were first counted and recorded the correlation between sunspot numbers and earth's surface temperature has been very strong. The brief period at the beginning of this cycle was used in a very unscientific way by advocates of global warming who are now being proven wrong.

Additional Authors Note: Most of you are aware that the global warming doctrine maintains that we are in a period of increasingly rapid increases in the earth's surface temperature. We will discuss in Chapters 9, and 10 how their data manipulation, (which includes admitted changing of the data they use, and other deceptive techniques) are used to make that claim. The most reliable unprocessed climatic data indicates that the real temperature rise has been between 0.45⁰ to 0.55⁰C. Earth's surface temperature has remained static or decreased slightly during the last 25 years. (Note: This statement does not recognize, as legitimate, the vast numbers of raw data changes, homogenization of data sets, increasing the numbers of observation sites, data smoothing, long term rolling averages and changes in the base period, used to "hide the dip" See: Chapter 9, and 10 and <fatalflaws.net> page on Evolution of the Global Anomaly.)

We now have 44 years of precise satellite data that proves the correlation is strong. The physical mechanism has been monitored and tested by 3 different satellite systems, but global warming disciples still insist that all warming of the earth's surface over the last 140 years is due to increasing atmospheric CO_2 and other minor greenhouse gasses.

In addition to the above, the graph of sunspot activity on p. 26 is provided as general information and to illustrate two salient points, which are:

1. For some who are old enough to have lived during that time, notice solar cycle 20. The latter portion of this cycle and the early portion of cycle 21 covers the 1970s to 1979. During that time period one of the frequent topics in science, magazines and news articles was the question; "Are we entering a new ice age"? During this period the Arctic ice cap expanded significantly, providing the base line for Arctic ice extent (1980), that is now considered to be the "normal". Contemplate what that means.

Does that sound like what you would expect from "science," if that fact is never considered or disclosed?

2. Note that in 2004 the waning portion of cycle 23 dropped below 100 sunspots per month and cycle 24 barely exceeded this threshold. This period was used to challenge the correlation between earth's temperature and sunspot activity as mentioned above. As you can see, cycle 24 has been much weaker than cycle 20 and the other cycles (19, 21, 22, 23) have been the strongest in history. It is no surprise that the earth has been warming since the 1940s. Note: According to NASA the Mean Global Surface Temperature anomaly for 2016 was 0.96^{0}C. Of that amount approximately 50% results from changes and manipulation of the data as will be discussed later.

Authors Note: Technically the raw data is not being changed. It is still there, but no one uses it. The raw data is used to generate the "unadjusted data" in a 5 step "correction process". It is this "unadjusted data" that is then used to determine the anomaly and other analysis procedures. Yes, it is a very misleading claim when NASA or anyone else says the raw data is not changed. It is a truth, used to mask a lie.

A Rhetorical question: Why is it that a weak positive correlation between earth's surface temperature and increasing CO_2 content of the air during the last 140 years proves the global warming doctrine. But, five historical episodes of strong correlation between low sunspot numbers and cooling of the earth is not proof of a relationship between solar activity and earth's surface temperature?

We are supposed to believe (have blind faith) that the very high level of solar activity over the last 78 years has had no impact on the earth's surface temperature. Consider the following; This supports what I have been saying.

"We are currently in one of the deepest solar minimum periods seen in the space age" see <https://spaceweatherarchive.com/2018/09/27/the-chill-of-solar-minimum/>.

"Sunspots have been absent for most of 2018, after going through the normal solar cycle decline period before 2018". According to Dr. Tony Phillips (see above link): "We see a cooling trend," says Martin Mlynczak of NASA's Langley Research Center. "High above Earth's surface, near the edge of space, our atmosphere is losing heat energy. If current trends continue, it could soon set a Space Age record for cold." Note that these measurements are of emissions originating 100 to 300 Kilometers above the earth. They are measuring the variation in the glow (irradiance emissions) of Nitric Oxide and CO_2 which increases as solar irradiance levels increase and decreases as irradiance declines.

Close to the end of this article is a graph of the Thermosphere Climate Index. Note how closely this graph reflects the sunspot number graph on p. 26. Cycles 19,21,22 and 23 were historically high in their sunspot numbers. Cycles 20,24 and now 25 have been significantly lower.

What is the cause of this cooling of the outer atmosphere? Quoting from the first paragraph of the above-mentioned report: "The sun is entering one of the deepest Solar Minima of the Space Age. Sunspots have been absent for most of 2018, and the sun's ultraviolet output has sharply dropped. New research shows that Earth's upper atmosphere is responding."

Quoting further: "These results come from the SABER instrument onboard NASA's TIMED satellite. SABER monitors infrared emissions from CO_2 and nitric oxide, two substances that play a key role in the energy balance of air 100 to 300 kilometers above our planet's surface. By measuring the infrared glow of these molecules, SABER can assess the thermal state of gas at the very top of the atmosphere–a layer researchers call "the thermosphere."

"The thermosphere always cools off during Solar Minimum. It's one of the most important ways the solar cycle affects our planet," explains Mlynczak, who is the associate principal investigator for SABER.

Author's Note: The above mentioned, deep solar minimum has extended through all of 2019, 2020, to February of 2021 when the sunspot numbers began to increase in cycle 25. You can follow what is happening on a daily basis at: www.spaceweather.com.

Additional Authors Note: At the end of the above mentioned article there is a link that will take you to a critique of claims that the cooling of the Thermosphere will result in cooling of the earth's surface and near earth atmosphere. That link is dated 2018 as is the information therein. It is in direct conflict with the NASA statement quoted on pp. 27-28. The assumption is made that only the thermosphere has been influenced by the lower levels of solar radiation (an invalid assumption). This is an example of how global warming doctrine projects uni-dimensional, current information onto a very complex system and draws conclusions about future occurrences in a very unscientific way. Let me explain.

As solar radiant energy moves through the atmosphere some of it is absorbed by various atmospheric components. The deeper into the atmosphere that radiant energy moves, the denser the various components become. At sea level the barometric pressure is approximately 15 pounds per square inch. At 10,000 feet the barometric pressure is approximately 10 pounds per square inch. This means that the atoms of the various elements that make up the atmosphere are further apart (less dense) as altitude increases.

Approximately 80% of the atmosphere (Nitrogen, oxygen, Aragon, and minor greenhouse gasses) is in the lower 40,000 feet. This is the area where most of the weather and climate functions take place (and where a high percentage of the water vapor is). The above mentioned SABER instrument is monitoring the temperature of CO^2 and NO at the 100 to 300 kilometer altitude. This instrument "sees" the molecules of these atmospheric components that have absorbed infrared energy which excites, or heats them. When they cool, some of the absorbed energy is radiated back into space. The higher the altitude where this energy is absorbed, the higher the percentage of that energy will be radiated into space. (See Appendix 2: p. 267)

This absorption and re-radiation of solar energy takes place from the top of the atmosphere to the earth's surface. In this process the amount of solar radiant energy that reaches the earth's surface is decreased. It is the opposite of the greenhouse effect. It results in the earth's surface being cooler. As the concentration of CO_2 in the atmosphere has increased over the last 140 years, this cooling effect has increased.

Rhetorical Questions: How much of the contribution that rising CO_2 levels have made to the greenhouse effect, has been offset by the above mentioned cooling effect? Has this cooling impact been investigated, and reported by global warming advocates? What is the net effect of the 121 ppm rise in the CO_2 concentration?

On page 267 you will find Appendix #2 entitled: "The Ultimate Fatal Fail of the Global Warming/Climate Change Theory"; There you will find my answer to the above rhetorical questions. It will probably shock some and surprise many, who recognize the truthfulness of it.

Continuing from p. 32: Quoting Martin Mlynczak of NASA's Langley Research Center. "The sun is entering one of the deepest Solar Minima of the Space Age. Sunspots have been absent for most of 2018, and the sun's ultraviolet output has sharply dropped. New research shows that Earth's upper atmosphere is responding."

The above quote and many other sources are unanimous in their assertions that as sunspot numbers decrease, the solar irradiance that reaches the outer atmosphere of earth decreases ("the sun's ultraviolet output has sharply dropped"). The decrease in solar irradiance wanes across the entire spectrum; from X-rays, ultra-violet, the visible and the infrared spectrum.

The reverse is also true, as the solar sunspot and solar flare activity increases there is an increase in the suns surface temperature which means more solar energy reaches the earth's outer atmosphere and the earth's surface. For an authoritative summary of the causes, effects, and magnitude of the variation in the solar irradiance that reaches earth see: <https://earthobservatory.nasa.gov/features/SORCE/sorce_03.php> Solar Variability under the heading Solar Radiation and Climate Experiment (SORCE)

The above mentioned impact is magnified when one considers that much of the potential greenhouse effect of increasing CO_2 is by-passed by approximately 75% of the heat which leaves the earth's surface as potential chemical energy, due to evaporation of water.

The relatively small variations in the sunlight that reaches earth's surface, interact with water in ways that actually increases its impact on the movement of heat from the earth's surface into the mid levels of the atmosphere. This type of synergistic interaction is completely ignored by global warming dogma even though, as we will discuss, it is extremely significant.

Recently, additional information has become available from detailed studies which looked at the solar energy given off by the "quiet sun," or the sun when there are no sunspots or flares (as illustrated by the photo series on p. 25; note the darker appearance of the sun in 1985 and 86). The graph below illustrates how the energy released from the quiet sun has increased over time since 1910. The red line shows the solar irradiance that reached the earth's outer atmosphere. Up to the mid 1970s the values were predicted based on sunspot numbers and electromagnetic, X-ray and UV energy records. Since the 1970s, the lines represent the satellite instrument measured solar radiation.

In the following paragraphs you will see expressions of sunlight intensity in terms of W/ m^2. This means watts per square meter at an elevation of 200 miles above the earth's surface. It may be a little confusing for some, but the short message is; the sun has gotten hotter because of solar activity since 1890 (seen in the graph below as the increasing height of the bases of the peaks). The extra energy reaching the earth, due to the increase in the energy released by the quiet sun and the high level of sunspot activity is sufficient to account for approximately 50% of the earth's surface temperature increase. This is not a human caused impact, it is natural variation.

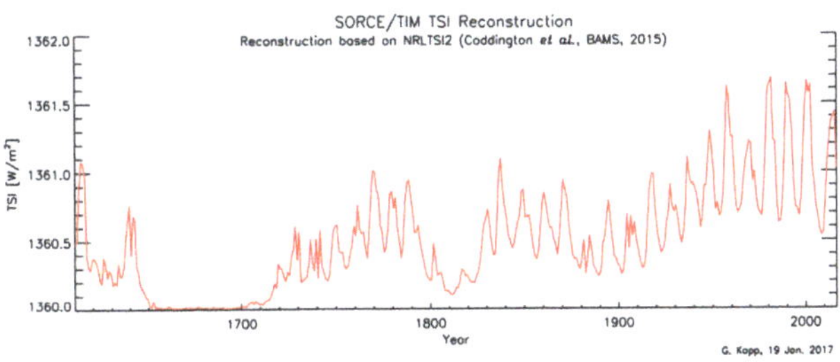

The following explains the scientific basis for the statement above. Based on the actual measurements of solar energy at the quiet sun stage of each solar cycle over the last 48 years two different studies concluded that; since the Dalton Minimum (the low between 1790 to 1830) the energy output of the quiet sun has increased by 0.04% to 0.16%. This range represents 0.544 W/m^2 to 2.176 W/m^2. (The figures in the last sentence represents the conclusions of two different studies.)

In the following discussion I will use the lower of the two figures (0.544 W/m^2 from the above indicated graph source) because it is more widely accepted and agrees with estimates I have made using different methods.

The solar energy at the solar maxima has also increased by approximately 0.7 W/m^2 averaged over the last 6 solar cycles. This means that the active portion of the last 6 solar cycles has averaged a 0.35 W/m^2 increase in solar irradiance that reaches the earth's atmosphere.

When the increase in the solar irradiance from the quiet sun and the increase from the active portion of the solar cycle are added the number becomes (0.544W/m^2 + 0.35W/m^2 = 0.894) 0.894W/m^2 at the top of the earth's outer atmosphere.

NASA and the UNIPCC agree that 24% of the solar energy that reaches the earth's outer atmosphere is absorbed by the earth's surface. This means that the study discussed above indicates that 0.215 W/m^2 of increased solar irradiance is absorbed (retained) by the earth's surface now compared to 1890.

This is a highly significant warming impact representing 25.3% of the excess energy that NASA and the UNIPCC says has been absorbed by the earth's surface. (0.215W/m^2 / 0.85W/m^2 =

25.3%) If the effect of data manipulation is taken into account, this figure, 25.3%, becomes 50.6%.

Author's Note: The solar radiant energy at the edge of the earth's atmosphere was estimated to have been 1360W/m^2 during the Maunder Minimum [1640-1710], 1360.11W/m^2 at the low point of the Dalton Minimum [1810] and 1360.32 W/m^2 in 1890. Using the 1890 figure you will notice that the base of the solar cycle events steadily increased until about 1965 when it reached 1360.7 W/m^2. It then stayed near that level until the end of cycle 23 in 2010.

This shows that the emissions that reach earth's atmosphere from the quiet sun and the solar cycle average, over the last 6 cycles, have increased by 0.38W/m^2 + 0.5W/m^2, for a total of 0.88Wm^2. This is a warming factor that is equal to 25% of the estimated excess heat that has caused earth's surface temperature to increase since 1890. This is the Author's calculations which are even more conservative than the most conservative reported by the authors of the research reported in the paragraphs above. If it is 23% or 25% that means that the core assumption of the Global Warming Doctrine is wrong (See <fatalflaws.net> Evolution of the Global Anomaly).

The information in the above note is consistent with the NASA information quoted above: (NASA earth observatory; Solar Radiation and Climate Experiment (SORCE); January 21, 2003) "Modern scientists believe that since this minimum (Maunder minimum) in solar energy output, there has been a slow increase in the overall sunspots and solar energy throughout each subsequent 11-year cycle.")

These totals are very close to my previous statements, even though the methods used in these studies took a very different approach than the method I used to reach the conclusions reported earlier as outlined in the Author's note above. That should be seen as strengthening the case for both methods and results.

This warming contribution does not include the warming that has resulted from ozone depletion. This phenomena permits a higher percentage of solar X-ray and ultraviolet light to reach the earth's surface, resulting in another significant warming force (See p. 42 for supporting research).

Neither does it include the warming that has resulted from the ElNino cycle as reported by Lawrence Livermore National Laboratories in November of 2018.

The suns increasing energy output has made a major contribution to the warming of the earth's surface during the last 78 years. There has also been the contribution of the two warming factors mentioned in the paragraphs immediately above. These contributions have been erroneously ascribed to the greenhouse effect of carbon dioxide and other minor greenhouse gases. This fact constitutes a major fatal fail of the global warming/climate change dogma.

There is no doubt that the sun is the most important factor in determining the earth's surface temperature. There is strong evidence that throughout the last 400+ years the earth's surface temperature has varied in response to the suns solar energy output.

Thanks to modern technological advances we have been able to monitor the solar activity and the resultant changes in the amount of solar energy that reaches the outer limits of earth's atmosphere. The correlation between the heating impact of sunspot numbers, ozone depletion and the ElNino effect on earth's surface temperature is very strong. These impacts all meet the requirements of Jacobs Rules to qualify as causative warming forces proving that the Global Warming Theory is fatally flawed. (See: A Correlation Primer p. 253)

We examined a chart of the sunspot and solar flare numbers from 1610 to 2014 and were able to see the solar maximum and minimums, the prolonged minimums, and the marked increase in the average solar activity that occurred between 1940 and the present. All this basically proves, that variations in solar activity have been a major part of the small but significant increase in the earth's surface temperature.

Author's Note: I recently observed a chart showing the LOWESS smoothed comparison between sunspot numbers and the earth's surface temperature for the time period of 1950 to present. The smoothed line incorporated a 11 year rolling average and showed a negative correlation between the sunspot number and the surface temperature of the earth. It "proved" the opposite of what I have indicated. The average sunspots decreased during a period when the temperature was increasing. This chart constitutes a total deception, brought about by the application of a highly suspect statistical technique (trick), combined with a carefully chosen "start date" during the highest period of solar activity in history. (See pp. 232 - 237 for a perfect example of how LOWESS smoothing changes a result.)

Take a look at the graphs on pages 21, 26 and 35, noting how the solar cycles from the 1940s to the current are much more intense than any period in history. The Encyclopedia Britannica chart on page 21 is based on annual averages, while the Royal Dutch observatory chart is based on the monthly average sunspot numbers. The chart on p. 26 is much more detailed. In it you can see how cycle 19 (1950 to 1964) was, by a significant margin, the most active in history, with 4 years of monthly average sunspot numbers over 200 with the peak reaching 350. Cycle 20 was not nearly as active if you compare it to cycles 19, 21,22 and 23, but was still more active than any cycle between 1610 and 1930. So if you want to "prove a point" based on a lie, use LOWESS smoothing with a long term rolling average comparison between sunspot number and temperature using a start date of 1950.

Let me repeat the point made in the last paragraph. What this plot proved is that you have to be very careful about anyone who uses LOWESS smoothing. It can be used to mislead, and often is. Cycle 19 (1950-1964) was the most active in terms of average monthly numbers of sunspots in recorded history. Subsequent cycles have had lower maxima but those maxima have all been higher than any cycle between 1620 and 1940. You will see additional examples of how LOWESS smoothing changes results and misleads in Chapter 10.

Update 10/28/2019 The following from: Weather.gov > News Around NOAA > "Solar experts predict the Sun's activity in Solar Cycle 25 to be below average, similar to Solar Cycle 24." If this trend continues the probability is high that the average global mean temperature anomaly will continue the downward trend of 2017 & 2018 which showed a 16% decrease in the anomaly. (When I wrote this paragraph initially I was not aware that NOAA and other global warming advocates were in the process of adopting, a new data set which includes 13,000 additional observation sites {The "Mann Trick" we will discuss later} to prevent the continued downward trend in the anomaly.)

You will find significant differences in historical numbers at different sites because of the way counting varies and due to LOWESS smoothing. As a reference I would suggest that the TIM graph displayed on p. 35 is state of the art and not yet subject to data manipulation.

Recently I happened across an authoritative source that addresses a question I have had for a number of years. That question concerns ozone depletion, which has been a major area of concern, beginning in the mid-1970s. If you are old enough, you will remember the southern hemisphere "Ozone Hole".

The concern was that the ozone in the upper atmosphere which plays a major role in limiting the level of ultraviolet and X-ray radiation reaching earth's surface was being depleted. These wave-lengths are "very high energy" emissions. One photon {packet of energy} at these wavelengths contains several times the energy of a photon of infrared energy.

At: <https://apam.columbia.edu/lorenzo-polvani> you will find background and information concerning Dr. Polvani who is a Geophysicist at Columbia University. Read his background information at this site and then you be the judge of his (and his co-authors) conclusions. Particularly pay attention to the following description of his interests and research: "He is particularly interested in quantifying and distinguishing the effects of anthropogenic forcing (human caused climate drivers) from natural climate variability." For a deep dive into his work you can access a number of his scientific writings by clicking on the "Publications" next to his picture.

Several of these articles show that things other than CO_2 forcing of the atmosphere have contributed to the rise in the earth's surface temperature.

Briefly here is a description of conclusions as to the impact of decreased ozone levels in the upper atmosphere. The reader is encouraged to go to: <http://www.columbia.edu/~lmp/paps/polvani+etal-NATURECC-2020.pdf> and read the Abstract (or preferably the entire article).

1. The damage to the ozone levels in the upper atmosphere which was caused by CFC's (a class of refrigerants) was the primary cause of the "ozone hole" over the Antarctic.

2. The ozone destruction was most dramatic over the southern pole area, but was significant over the entire globe.

3. Ozone-depleting substances (ODS) have enabled a higher percentage of the solar energy that arrives at the outer limits of the earth's atmosphere to reach the earth's surface (a warming impact).

4. Quoting from the abstract of a Letter published on 20 Jan 2020 in the journal Nature/ Climate Change "We also demonstrate that the large impact of ODS on the Arctic occurs primarily via direct radiative warming, not via ozone depletion".

The sentence immediately above translated into lay terms says that the impacts they measured were the effect of increased sunlight directly warming earth's surface, not through greenhouse impact enhancement.

5. The abstract continues "Our findings reveal a substantial contribution of ODS to recent Arctic warming, and highlight the importance of the Montreal Protocol as a major climate change-mitigation treaty".

6. The above quote refers to the "Montreal Protocol" which has been in place as a world wide treaty since 1987. This treaty phases out the use of ozone depleting substances (ODS). This phase-out was to be complete by 2005 and that has allowed the ozone in the atmosphere to begin the process of reestablishment. A portion of the depleted ozone has been naturally re-

established but it will take years before it is back to the early 1900s level.

The bottom line is that ozone depletion, a human caused effect, has been a major contributor to the warming of the earth's surface over the last 70+ years. It has been effectively limited and reversed as long as the world community is willing to enforce the terms of the agreements. Lower CO_2 concentrations in the atmosphere would have done nothing to lessen the effects of this warming force.

Chapter Summary:

The sunlight irradiance reaching the earth's surface over the last 78 years has been historically high because of 3 different factors.

First: The increased level of solar activity beginning in the early 1900s followed by a dramatic increase with solar cycle 19 has resulted in higher levels of solar irradiance reaching earth's surface. Also the increasing temperature of the sun during the quiet stage of the solar cycles has made a significant contribution.

Second: The Lawrence Livermore Laboratory study showed that 40% to 50% of the Arctic ice loss over the last 38 years, was the result of natural variation in weather patterns (not human caused).

ElNino causes unusual high pressure systems around the North and South Poles. These high pressure areas, with their attendant low cloud cover conditions, allows much more solar radiation to reach the earth's surface. This is important because of the long day lengths during the "summer" across the Arctic and Antarctic.

Third: The increased sunlight which has been reaching earth's surface, because of ODS activity, has played a significant role in the warming since 1950 as outlined above. (This impact is now decreasing because of science based change, driven by international treaty. It is human caused but not related to the greenhouse effect).

The three impacts delineated above illustrate how the basic controlling assumption of the global warming theory results in it fatally failing. All of the earth's surface warming during the last 140 years has not been the result of CO_2 forcing! It is no surprise that the sun has been responsible for a major portion of the earth's warming.

Any one of the three factors discussed above destroys the foundational assumption of the global warming theory; that assumption is; 100% of all the warming of the earths surface and near earth atmosphere over the last 140 years has been due to CO_2 forcing of earth's surface temperature.

Authors Note: for those who doubt my claim that global warming advocates believe that 100% of the earth's warming has been caused by CO_2 and other minor greenhouse gasses: Listen to Mr. Gores speech before the January 2023 "World Economic Forum" where he refers, several times, to anything else as "B—-S—-" a new scientific term.

Photo p. 45: Aspen and other tree species play important roles in the removal of carbon dioxide, other minor greenhouse gasses and pollutants from the atmosphere. The end result can be and is, the longterm sequestration of vast quantities of carbon dioxide and other minor greenhouse gasses. Photosynthesis and related processes decrease the amount of human caused CO_2 buildup in the atmosphere from 40+ billion tons, to approximately 11.35 billion tons annually. We will discuss this and other mechanisms which remove CO_2 from the atmosphere later in this work.

Chapter 2: The Little Blue Marble Deception

This chapter is straight forward: No synopsis
No Summary.

In Mr. Gore's first two books on "Global Warming" he discussed in some detail how a dedicated scientist had used a huge number, (20,000) photos, of the earth, taken from space, and by meticulously removing the cloud cover had created a composite photo of the earth as it would be observed from space, if there were no clouds in the atmosphere.

He went on to opine that, this view of the earth demonstrated how delicate the environment of the "Little Blue Marble" is. This became a common expression repeated by politicians, climate scientists, news correspondents and even NASA officials and NASA on-line articles (and is being taught, as fact, in K-12 and our universities).

This idea; that the earth is small and delicate, became a major portion of the foundation of the global warming doctrine. I was deeply impressed when I first saw the above mentioned photo. When I read the conclusion reached by Mr. Gore, that this view of the earth demonstrated how delicate the earth's environment is, I was instantly aware of the deception involved. I also realized that few people who read those words would recognize that deception for what it is.

There are two fundamental problems with this concept: (I will reserve judgment as to whether or not the deception was intentional or not - - It could have been unintentional if the author lacked an understanding of a vast volume of the science involved). The problems with the concept are:

First: Viewing the earth from space, whether from 200 miles up, or from the moon does not change the size, mass, or nature of the earth.

a. The circumference of the earth is 40,075 km (roughly equal to 24,901miles). It is not small.

b. The volume of the earth is $108.32X10^{10}$ km^3 (1,083,200,000,000 cubic kilometers). It is not small.

c. The mass of the atmosphere is 5,148,000,000,000,000,000 kg, or 11,325,560,000,000,000,000 pounds. This is the mass of "dry air" and does not include the mass of liquid water, ice, and water vapor. It is not small.

d. The mean mass of the water vapor in the atmosphere is $1.27X10^{16}$kg (12,700,000,000,000,000 kg or 27,940,000,000,000,000 pounds) when the increase in the mass is adjusted by the $0.5x10^{15}$kg — (that is 1,100 Trillion pounds) of additional mass of water vapor present during E*l-Nino* periods. That is not a small amount.

e. At any one moment the amount of water contained in the atmosphere in the liquid and solid states (clouds) is roughly 10 times the amount of water vapor (equals 279,400,000,000,000,000 pounds). That is not a small amount.

f. The earth is not small, but is a vast, complex combination of interacting elements, substances and processes. The more one learns about it, the more intricate, complex, resilient and beautiful it appears.

Second: The earth has never existed in the "little blue marble" state. That state is an artifact, resulting from meticulous effort to show the earth as it has never existed. Contemplate:

a. The earth's surface is composed of approximately 72% ocean surface, plus, the surface areas covered by rivers, streams, lakes, glaciers, wetlands and ice deposits. The extensive water surface area plays a major role (second only to the sun) in climate control and modulation as well as buffering changes in the CO_2 concentration. It also leads to the sequestration of huge amounts of CO_2.

b. Water held in the soil, rocks, and plant communities also plays a major role in regulating and modulating the earth's environment.

c. Water vapor, liquid and ice in the atmosphere play a major role in limiting the amount of sunlight that reaches the earth's surface. Clouds reflect a significant percentage of the incoming sunlight back into space and absorb another significant amount of the solar energy, much of which is then radiated back into space. Without this function of the cloud cover and water vapor the temperature of the earth would be considerably higher than it is now.

Mr. Gore dismissed the role that water, water vapor, and ice plays in regulating the earth's temperature because it varies widely, changes rapidly, and is difficult to quantify. We know a great deal about water; that knowledge deserves much more than a glib dismissal. True science is seldom easy.

In chapter 3, you will be introduced to some of the amazing physical characteristic of water. You will also learn why this substance is unlike any other on the face of the earth. The cycle of evaporation, ascendance into the mid atmosphere, followed by condensation and eventual precipitation is truly one of the miracles that we seldom think about and appreciate.

Photo: Water is a miraculous substance, because of its unique physical and ascetic properties. In settings like the one shown above, it also makes a wonderful home for Bonneville Cut Throat Trout. I am an ex-fisherman; but, if one of my grandchildren wants to learn to fish I can quickly revert to prior habits. For more on how water buffers and controls the climate visit: <fatalflaws.net>

Chapter 3: Water's Role in Controlling and Modulating the Earth's Surface Temperature

Synopsis: *Water is the second most important factor in the control of the earth's surface temperature, exceeded only by the sun and its influence. It is both a feedback (buffer against temperature change), a role acknowledged by global warming advocates, and a primary driver of weather and climate, a role ignored, even denied by the disciples of the doctrine.*

When Mr. Gore glibly dismissed the role of water in his first books on global warming, he established "the word" for global warming/climate change disciples which has been accepted on blind faith. It could not be more erroneous. In dismissing water's contribution to the control of earth's surface temperature and climate he did two things.

First; He guaranteed that the global warming movement would be established on a foundation of sand (or chose your own metaphor from jello, quick sand, etc).

Second; He assured, that any warming influence that water contributes, would be credited to CO_2 forcing of earth's temperature. An example of how that happens was included in the Book Abstract when we discussed how 40 to 50% of all the ice loss from the Arctic ice pack has been caused by natural climatic variation traceable to ElNino effects. ElNino is a sea surface temperature driven phenomena. It is part of what scientists from Lawrence Livermore National Laboratories refer to as natural variation, i.e., it is not human caused.

How many times have you heard the apocalyptic pronouncement that; "The Arctic Ice is Melting, and it is because of rising atmospheric carbon dioxide". Unless you live deep in the woods with no "devices" a good estimate would be once a day for 30 years (equals 10,950 times).

It was not and is not true. We will discuss, later in this work additional causes of Arctic and Antarctic ice melt. That part of the scientific community who still have open minds are beginning to appreciate this fact.

Some of the extremely important roles water plays in the control and modulation of the earth's surface temperature are:

1. Water evaporates from the earth's surface, then rises into the mid levels of the atmosphere, because it is much lighter than the other atmospheric components. Due to the cooler temperatures of the mid-altitudes, it condenses into water droplets. This process transports approximately 75% of all the heat which leaves the earths surface into the mid levels of the atmosphere. Some of the heat released when this water condenses is radiated into space, and some warms the other gasses by convection or direct contact heat exchange. This process by-passes much of the potential greenhouse gas impact of CO_2*, methane and other minor greenhouse gases.*

2. Water vapor is much lighter than atmospheric nitrogen (N_2*), oxygen (*O_2*), and carbon dioxide (*CO_2*). Because of this physical/chemical characteristic of water vapor, it is the driving force behind changes in atmospheric pressure. Water vapor, or low levels there-of, causes the low pressure and high pressure systems, so important in determining the weather in the short term and climate in the long term. (In some publications you will find scientists and NASA officials saying that the movement of water vapor in the* *atmosphere is random. That is not the case. Because it is much lighter than other atmospheric components, the movement is upward, just like the helium filled balloon my grandson let slip from his fingers.)*

3. The concentration of water vapor molecules in the atmosphere is at least 24.4 times that of CO_2. When the liquid and solid state water in the atmosphere is added, the concentration of water dwarf's the concentration of carbon dioxide. In addition, the liquid and solid state water in the atmosphere increases the effectiveness of water in controlling the greenhouse effect to the point that the minor greenhouse gasses role is almost negated.

The above three facts constitute the second most important fatal fail of the global warming/climate change theory. All this and more will be discussed in this chapter.

Chapter 3: Water is the most widely distributed substance on the earth's surface. Seas cover approximately 72% of the earth. When you add to that, the surfaces of lakes, streams, wetlands, and plant communities that transpire water, the figure is much higher.

When the wide distribution of water is combined with its total mass and amazing physical properties it dwarfs the effects of all other substances in the control and modulation of earth's environment.

If you want to understand the truth about global warming you must understand water's roles in the environment, and that its net effect on warming of earth's surface, is not equal to zero, as is assumed by the global warming doctrine.

I will try to avoid boring numbers; there will be some but just look at them and classify them into one of four categories, which are: 1= Tiny, 2= Large, 3= Huge, 4= I can't even say that one.

Unlike CO_2, methane, ozone, nitrous oxide and other minor greenhouse gases, water should be referred to as a greenhouse substance, because it exists in all three physical states (gas, liquid, solid) in the atmosphere. This means that water is capable of absorbing a much broader spectrum ('color' range) of sunlight and radiation given off by the earth, compared to the minor greenhouse gases. It is a much stronger greenhouse agent and in much greater concentrations in the atmosphere.

An Example: According to USU Climate Center data, on August 22, 2018, between 11 am and 4:00 pm at Cherry Hill Farms – Santaquin UT west site (an automated, multi-function weather station); the total solar radiant energy intensity at 10 feet above ground level varied between 1096 to 1125 W/m^2 (Watts per square meter) under clear skies. A very large and intense thunderstorm moved into the area at 4:00 pm and the light intensity dropped to less than 17 W/m^2, equal to very late twilight levels, within 15 minutes.

This data simply describes in numbers what you know. When clouds get between you and the sun, much of the light is reflected back into space, or absorbed by the water in the cloud. There is water vapor you can't see, thin clouds and dark clouds. Depending on how much water is between you and the sun, the sunlight you are exposed to changes. Rhetorical Question: Have you ever "enjoyed the shade" created by the four hundredths of 1% of the atmosphere that is CO_2?

(Please Excuse a technical moment) The weight of water vapor in the atmosphere is approximately 10 times (H_2O = 0.4%; CO_2 = 0.04%) the weight of CO_2 during non-El Nino periods. During El Nino the weight of water in the atmosphere increases by 1,100 Trillion pounds (that is 1,100,000,000,000,000 pounds [a #4 number]). One molecule of CO_2 weights 2.44 times as much as one molecule of water. This means that there are 24.4 times as many water vapor molecules in the atmosphere as CO_2 molecules. Add to that the liquid and solid water in the atmosphere and you can begin to get a feel for the magnitude of the difference between these substances.

With the above "technical moment" in mind do you realize that the global warming/climate change doctrine assumes that variation in water content of the atmosphere has no net effect on global surface temperature? Even the extra 1,100 Trillion pounds added to the atmosphere during El Nino. Astounding isn't it? Or is it a foolish assumption? Does the Lawrence Livermore National Laboratories article discussed earlier prove it is a foolish assumption? In the teenage vernacular - DA! *(The preceding DA does not include the impact of the North Atlantic oscillation which is driven by the sea surface temperature of the Atlantic around the Azores vs the temperature of the North Atlantic around Iceland and Svalbard which has the same climate driving characteristics as ElNino/LaNino. There are similar patterns in the southern hemisphere.)*

Water has the highest specific heat of any common substance. Don't let the technical appearance of that sentence turn you off. Think of the last time you placed a pan of water on the stove. It seemed like it took a long time for the water to boil. That is because, to heat water, requires a lot of energy.

When you consider the quantity of water in the earth's atmosphere it is evident that, to increase the temperature of that water by just one degree, would take a tremendous amount of energy. The number of calories required to do that is

15,260,000,000,000,000 or 15,260 Trillion calories. I will not bore you with what it would take to change the temperature of all the water on earth, let us agree that it would be beyond a #4 number. This is important because it means that water buffers the earth's environment against temperature changes.

Next, let us examine what the latent heat of vaporization {evaporation} is, and why it is extremely important.

So that I don't lose those of you who don't like boring scientific discussions I will ask you to preform an experiment in four easy steps. 1st. Place your index finger in your mouth (you may wash it first, if you would like). 2nd. Remove your index finger and place it approximately four inches from your mouth. 3rd. Blow hard on your finger. 4th. Note the drop in the surface temperature (cooling) of your finger. You now understand what the latent heat of vaporization is. A science type would define it as the amount of heat required to evaporate one gram of water (512 calories per gram of water), but you and I know what it is, right?

I can hear you asking: This is important; Why? That is an easy question to answer. It is estimated that each day 2,794 Trillion pounds of water evaporate from earth's surfaces (that is 2,794,000,000,000,000 pounds {a #4 number}). Each gram of water requires the input of 512 calories (about the calories in a Quarter Pounder) to convert it from the liquid to vapor state. (Did I lose any of you at the 2,794 Trillion Quarter Pounders?).

If you are still with me, I am sure you remember there are 454 grams per pound, so the total calories required to evaporate one days water from the earth's surface is equal to 1,268,476

Trillion Quarter Pounders (that is a huge pile of Quarter Pounders). Even if I lost you, hang on because what happens next is important and exciting if you want the truth about global warming.

When water evaporates it rises, taking the latent heat of vaporization into progressively higher and higher layers of the atmosphere. The energy is in chemical potential energy form, not as radiant energy. This means that this energy is not subject to interception by greenhouse gasses. It by-passes much of the potential greenhouse gas effect, on its way to an altitude where it will condense into the liquid state, because of the decreasing temperatures associated with higher altitudes

Authors Note: Under warm or hot conditions, the small droplets of water are carried along to even higher altitudes by the thermal effects of the mass of upward moving water vapor. This carries most of the heat released by condensation to even higher altitudes. This is what powers thunderstorms, hurricanes, typhoons and hailstorms.

Let me illustrate that last point. Several years ago I found myself at the church approximately six miles from home, at night, with no ride (that can happen when you have three teenage sons). As I left the Church, intending to walk home a friend saw me, (he was on his way to his deer hunting camp) and upon learning the situation offered me a ride. I accepted and he took me to the front gate, which is one mile from home, and dropped me off (without a car, I had no electronic key and could not open the gate). I enjoyed hiking the last mile which was much easier than hiking six miles.

In the above analogy, the first five miles of hiking were by-passed by the ride offered and accepted. When water evaporates, the energy it carries by-passes the layers of air close to earth and is transported upward to the level where it condenses.

At that point the heat of vaporization is released (dropped at the gate). It can then be radiated into space much easier than it could be from the earth's surface, having by-passed much of the potential greenhouse effect. The liquid or solid water (clouds), then becomes the major part of the greenhouse effect, dwarfing the impact of the minor greenhouse gases.

Authors note: When energy leaves the earth's surface in the form of radiant energy, that irradiance can be intercepted by any of the greenhouse gasses which absorb energy of the frequency (energy level) it has. It will then be re-radiated, again some toward space, some toward earth.

A single quanta/photon of energy may go through the absorption and re-radiation process several times before it returns to the earth's surface or is radiated into space. As the heat moves through these absorptions and re-radiations it may give off part of it's energy by direct contact exchange to other atmospheric components (a process called convective heat transfer). The radiant energy that is then given off as irradiance will have a longer wavelength (less energy). This, more or less random movement of energy tends to compound the greenhouse effect, particularly close to the earth's surface.

As the energy moves higher in the atmosphere, for example, to the 10,000 foot elevation, the atmosphere is one-third less dense, so the pathway to space is more direct. At between 16,000 and 19,000 foot elevation, the molecular spacing is increased so that the majority of irradiance released toward space exits the atmosphere. Also at this altitude the portion of the radiation released toward the earth is less and is more likely to be intercepted and re-radiated: some toward space and sometimes toward the earth.

Most of the condensation of the water occurs between 5,000' and 30,000'. The probability that it will be radiated into space is much higher than if it were radiated from the earth's surface. This is how, much of the heat, that is lost from the earth's surface, avoids most of the potential greenhouse effect.

This mechanism greatly decreases the impact of the very low concentrations of CO_2 and other minor greenhouse gases. Now that you understand that, you can understand how exaggerated the claims made about CO_2 forcing of the earth's surface temperatures are.

Authors Note: the author realizes he has been repetitious, it was intentional, you now understand why. This is a mechanism of heat loss from earth's surface that is totally ignored by global warming disciples.

The dew point of water is another physical property of this miraculous substance which plays a very important role in controlling the earth's surface temperature. The dew point is the temperature at which, the relative humidity of the air, as it cools, will reach 100% and any further decrease in temperature will result in formation of dew or frost on environmental surfaces (or water droplets in the air).

(Please Excuse another technical moment; I wouldn't do this if it was not important.) Remember, according to global warming dogma this process has no effect on the heating of the earth's surface. Water is a polar substance, with each molecule having a negative end (the oxygen atom) and a positive end (where the hydrogen atoms are connected to the oxygen). There are also weaker bonds known as Van Der Waals Forces that are shared bonds between molecules. The polar bonds and Van Der Waals forces are what gives water vapor the ability to exist as liquid and ice within the temperature range where life can exist (another reason water is a miraculous substance).

The preceding is not difficult to understand if you realize that there are three bonding forces involved. These multiple bond types, give water molecules a greater range of the energy spectrum that they may absorb radiant energy from. The chemical bonds between oxygen and hydrogen could be thought of as how a young couple hug each other, very strong and tight.

The second type of bonds involved, are the polar bonds between the positive and negative charged ends of the molecules. These bonds could be thought of as how you hug your in-laws or in-laws to be (cheek to cheek but little contact).

The third type of bonds are called Van Der Waals forces. These bonds could be thought of as how I hug my sister (shoulder to shoulder and hand resting on the opposite shoulder). Carbon Dioxide is non polar, it does not hug anything until the temperature decreases to -105^{0}C, then it is cold and hard (dry ice).

Part of the above is a little difficult to understand but it is critically important. Each of the three types of bonds will absorb, different wavelengths of energy.

As the air temperature cools, the association of water molecules with other water molecules increases. As water molecules bond together heat is given off, so as the dew point is approached the rate of temperature fall decreases (a warming effect). In addition, the association of water molecules and formation of liquid or ice spreads the absorption spectra of atmospheric water (a greenhouse effect enhancing factor).

When dew forms, the amount of heat given off is equal to the latent heat of vaporization; It is the reverse process of vaporization (remember the

Quarter Pounders) If frost is formed an additional amount of heat (the latent heat of fusion) is given off. This amount of heat is equal to a few fries and accompanies the burger, if the temperature is at or below the freezing point.

The following is a description of how the dew point of water helps regulate the earth's surface temperature. There are two examples of how the process works (I could give many more; be thankful for small favors).

First: I spend a lot of time hiking and exploring mountainous areas. What I experience there, occurs over much of earth's surface, even at times, desert areas.

As sundown approaches the temperature stabilizes, and then begins to fall. During mid-summer at 9,000 to 11,000 foot elevations on a bright sunny day the temperature may be 85 to 90 degrees one hour before sundown. When the sun goes down the rate of temperature fall increases and as it grows dark I begin to look for my jacket.

If I am still out at one hour after it grows dark (it seems that the older I get, the more this occurs; maybe I am getting slower) the temperature is low enough that a third layer is desired.

At some point in the time between sunset and darkness the air begins to feel moist and by 10:00 pm to midnight, walking through grass or brush, my feet and legs begin to grow wet. My wet feet tell me that the dew point has been reached and I know that (at least during the summer) it is not going to get much colder.

As mentioned before, as the temperature falls and approaches the dew point the rate of temperature decrease slows. When the dew point is reached the condensation of liquid on environmental surfaces begins to release significant amounts of heat which slows the temperature decline even more. At the same time, the association of water molecules in the air actually spreads the absorption spectra so that more of the radiant energy given off by the earth is intercepted (a greenhouse enhancing impact).

Even in desert areas the same temperature pattern occurs, it is just delayed, in most cases, because the relative humidity is lower and it requires a greater temperature decrease before the dew point is approached. By the way, this pattern is not easily observed during a winter walk in the dark, those are just cold at the start, cold in the middle and cold at the end. This is because our bodies try to maintain a minimum temperature of 98+ degrees. That is a total of 66 degrees higher than the air at the freezing point.

Second: Over a water surface, as evaporation occurs, the humidity of the air immediately above the water increases, reaching the dew point for a short distance (a few 1/16's of an inch, to a few feet). This higher water concentration in the air results in more of the radiant energy, which is released by the water surface, being trapped by the resulting greenhouse effect.

For several reasons, a liquid water surface does not give off as much radiant energy as a land surface but when the small amount that is radiated is multiplied by the total surface area it is significant. The higher water content of the air, immediately above the surface, suppresses the evaporation rate. Both of these processes slow heat loss from the surface and air layers immediately above that surface. Note: Air movement (wind) will quickly negate most of this impact.

This process slows the heat loss discussed in the section above which describes how the latent heat of vaporization moves huge amounts of heat from water or wet surfaces into the mid atmosphere and space. Without this very localized greenhouse effect, the rate of energy loss attributable to the above described mechanism would be greater than the 75% of all heat loss from the earth's surface, which would be significantly colder.

There are additional ways water helps modulate the earth's surface temperature both directly and by interactions with such forces as particulates, aerosols and heat from volcanic activity, magma flows and from the earth's mantel. These will be discussed later in the appropriate chapters.

The photo below is a small waterfall in Zion National Park, UT, USA. Water is also an important part of the beauty of many of the silent and magnificent parts of the earth. For billions of years it has carved, formed, and changed the earth in matchless ways. Water is also a wonderful solvent, and this physical property plays a number of important roles. One of the most important of these roles is that it dissolves large quantities of carbon dioxide, other gasses and pollutants from the air.

<u>Summary:</u> *Water is second only to the sun as a primary driver of the weather and climate of the earth. In Chapter 5 we will discuss how NASA and the Global Warming/Climate Change dogma considers water to be only a feedback mechanism. They wrongly assume, that it is not directly involved in the control of climatic patterns both regionally and globally.*

As is illustrated by peer reviewed, scientific publications such as the Lawrence Livermore National Laboratories study mentioned in this chapter and a vast number of other scientific findings. The quantity of water in the atmosphere does change, regionally and locally over vast areas of earth's surface. When it does change, the high and low pressure systems that drive storm, weather and climatic patterns also change.

Chapter 4: The Role of Heat From Earth's Core in Warming of the Earth's Surface Temperature

<u>Synopsis:</u> *If you have not been keeping track, this is chapter 4 and we are not to a discussion of how the minor greenhouse gases impact the earth's surface warming. I told you we would discuss the contributing factors in the sequence of their importance. We will get to* CO_2 *and other minor greenhouse gases in chapter 6, where they belong, in the order of importance in the control of earth's surface temperature.*

In this chapter we will discuss the mechanisms and degree of importance of how volcanic activity, magma flows, tectonic plate movement, heat rising from magma chambers within the earth's crust and heat rising from the earth's core contribute to the control of earth's surface temperature.

I realize that you can find many articles which claim to "debunk" the idea that volcanos are a significant factor in warming of the earth's surface. This line of thought ignores several very important factors.

First: the volcanic activity we see on earth is limited to the roughly 25% of the earth's surface that is covered by land.

Second: The number of volcanos under the seas, their size and activity is not considered in these "debunking articles" (it is not known). The earth's crust is thinner under the seas, so volcanic activity there may be more common than on land.

Third: In addition to the volcanos, there are huge amounts of heat that move to the earth's surface from magma flows on land and under the seas (Particularly along the tectonic plate margins; Magma intrusion is the force that drives plate movement. There are vast quantities of magma involved along these margins under the seas).

Fourth: Significant amounts of heat move by convection through the earth's mantle into the seas. This heat rises to the ocean surface in warmed water.

Fifth: Interactions between this heat source and the earth's waters, multiply its impacts.

Sixth: The locations of where this heat source rises to the surface is extremely important. For example, a volcano under the Arctic or Antarctic ice sheets can have both direct and indirect effects on the extent of the ice caps.

It is important that we remember several facts about how volcanic activity and heat from magma flows of various types impact the earth's surface temperature.

First: Volcanic eruptions (on land) propel large quantities of rock, particulate, sulfates and other substances into the atmosphere. These materials have a net cooling effect on the earth's surface by: a. they actually capture some of the energy of sunlight and re-radiate it back into space (a cooling effect), b. in addition, sulfates and aerosols are efficient in scattering light, much of which is scattered in the direction of space (a cooling effect), c. particulate may collect on reflective surfaces such as glaciers and ice packs which decreases their reflectivity (they absorb more of the suns radiant energy, a warming effect). The global warming models and earth surface energy budgets are fairly accurate in defining these cooling impacts. Most models ignore the warming impact.

Most of the effects mentioned above are dependent on the materials ejected into the atmosphere staying in suspension, only c is not dependent on the material staying suspended. Its' impact is actually enhanced as particulate settles too, and collects on environmental surfaces.

Particulates in the lower 40,000 feet of the atmosphere are removed fairly rapidly as they are hydrated by water, or serve as condensation nuclei for the formation of rain drops, snow flakes, or fog. This means that the impact of volcanic explosions directly on the atmosphere may be intense, but they tend to last a year or less.

Second: Volcanic activity close to, or under the oceans, impacts the earth's surface temperature in two ways. Volcanic materials ejected into the ocean water, or magma which is exuded along tectonic plate margins, into the water or onto the ocean floor has a very high temperature (1,000^0 to 2,500^{0}F or higher). This huge heat source *directly heats water. As water warms above 4^0 Celsius (39^0F) it becomes less dense. This means that it rises to the surface where it tends to spread over large areas.*

It must be remembered that, to the extent these heat sources increase the surface temperature of the earth's waters, the evaporation rate of water will be increased. Increasing the evaporation from earth's surface will transfer increased quantities of heat and water into the mid levels of the atmosphere.

This increased atmospheric water content will intensify waters contribution to the global greenhouse effect, further dwarfing the relative impact of the minor greenhouse gases. This is a wonderful and beautiful part of nature which has been modulating the earth's surface temperature for three plus billion years.

This is a great place to mention another of the fatal fails of the settled science status which has been declared to exist, pertaining to global warming/climate change dogma. Let us begin with four Rhetorical questions: I realize this is somewhat repetitive, but follow me and see if it is warranted.

First: If the underlying and supporting science behind global warming/climate change was settled, (settled defined as: disagreements resolved or agreement reached) how come there are multiple models that differ from one another in many ways?

Second: Why don't the different models treat climate driving or contributing forces the same way?

Third: Isn't it reasonable to expect all of the models to consider or deal with the same climatic driving or influencing factors?

Fourth: If the science has been settled from the time that it was declared to be, (at least 12 to 17 years ago, {the latter figure is the time that has lapsed since "An Inconvenient Truth" was published}), 1. Why have multiple data sets continued to be introduced and changed? 2. Why is raw data frequently changed? 3. Why are techniques such as homogenization of data sets, LOWESS smoothing and long term running averages of data sets required to make the data more doctrinally friendly? 4. Why is it common to run a model over and over until a significant result is reached, but not so many times that the error becomes unacceptable? That is not settled science. Is it science at all?

What the above four questions illustrate is internal inconsistency as well as a redefinition of what science is. Up to the time of the Paris meetings there had been five phases of assessments of a considerable number of different models. One participant mentioned that, of the 25 different models (he) assessed, only a handful (approximately 30%) addressed any contribution of volcanic activity to global warming. Those that did address a volcanic contribution only did so in terms of sulfates and volcanic ash's cooling impact as they intercept incoming solar radiation. If there was internal consistency concerning a settled issue the percentage should be 100% or 0%.

Chapter 4: The photo on p. 67 is of Nishinoshima Island ([Island of the East {english}] taken on October 18, 2015, by the Japanese Coast Guard. The photo shows the active lava flow emanating from the vent along with a gas plume and water discoloration. The island (in 2015) was approximately 1.9 x 1.9 km in size.

This island is the result of a volcanic eruption that broke the surface of the Pacific in 2013 and has been active since that time, with the exception of a few months from late 2017 to July of 2018. It is located near the junction of tectonic plates and the base is approximately 2,300 feet below the surface. In 1973 a new island emerged from the sea in this area and in 2013, in a spectacular eruption. This island emerged just south of the other (small earthquakes centered in this area have been frequent since the initial 1973 event). In situations of this type, earthquakes are usually associated with magma movement (heat source).

Authors note: The visible signs of volcanic activity have gradually waned over the last several years. The quakes in the area have continued, though the location (epicenter) is slightly different. As it turns out, recently it has been reported that another potential island is approaching the surface of the ocean a short distance to the southwest of Nishinoshima. Photo: Nishinoshima Island taken by the Japanese Coast Guard

This island is one of a number to emerge from the oceans over the last 14 years from the Samoan Islands, Indonesia, The Pacific (ring of fire), east of the Kamchatka Peninsula, Iceland and others.

When an event such as the emergence of an island through the sea surface is examined there are two physical impacts that are readily apparent. The first is the amount of material that is required to build the base and the height of the island upward through the surface. The second, when contemplated is equally important.

The material that builds the island comes from a magma chamber in the earth's mantel, at $1{,}000^0$ to $2{,}500^0$F. The heat carried by the material deposited, the steam given off, and the upwelling of heat carried and transmitted upward through the crust from the magma chamber is huge. The heat from deposited materials, and heat rising through the earth from the magma chamber warms vast quantities of water, which, because warm water expands, rises to the surface.

Additional authors note: (In the late development department): ***UNDERSEA VOLCANIC ERUPTION SEEN FROM SPACE:*** *An undersea volcano erupted on Jan. 15th, 2022, near the Pacific Island of Tonga. On shore, people rushed to higher ground. This volcano exploded with enough force to send volcanic debris over 60,000 feet into the atmosphere and send atmospheric shock waves that were seen in videos and recorded by barometers, as far away as Florida (Early reports indicate that this volcano is the strongest explosion in 1,100 years). More recent reports show that 400,000 tons of volcanic material were discharged into the stratosphere, with some of it reaching into the mesosphere.*

The above mentions only those volcanoes that have risen above the seas surface. With 75% of the earth's surface covered by oceans it would be expected that the numbers of active volcanos and lava flows that we don't see would be significant.

There is a third physical impact that occurs when volcanic activity or magma flows occur under or close to the polar ice caps. In this chapter we will discuss and cite examples of how volcanic activity has melted significant amounts of sea ice in the North Atlantic near Svalbard Norway. Also in the Northern Pacific along the coast from Hokkaido Japan northward into the gulf of Alaska and appears to be contributing to the melting or the West Antarctic and East Antarctic ice sheets.

To further assist in understanding this topic as it relates to earth's surface temperature let us consider for a moment examples of what happens below the sea surface and usually goes unnoticed.

Example #1: quoting from NBC News: https://www.nbcnews.com/mach/video/the-largest-underwater-volcano-explosion-the-world-almost-missed-1158176835781 "More than 70% of volcanic activity occurs underwater -- with much of it hidden from human view by seawater. But in 2012, passengers on a flight over the Pacific Ocean saw something odd: a big chunk of floating pumice rock, signaling an underwater volcano had erupted. A new paper published in the journal, Science Advances, details a 2015 exploration and its surprising discoveries."

How big was this volcano, and what was its impact? <https://www.sciencealert.com/almost-nobody-noticed-largest-underwater-volcano-eruption-ever-recorded-havre-seamount> In order to understand the magnitude of this event we need to see how it was described; Quoting; "We knew it was a large-scale eruption, approximately equivalent to the biggest eruption we've seen on land in the Twentieth—Century," says volcanologist Rebecca Carey from the University of Tasmania, who co-led the first close-up investigation of the historic 2012 eruption. "The incident, produced by a submarine volcano called the Havre Seamount, initially went unnoticed by scientists, but the floating rock platform it generated was harder to miss".

"Back in 2012, the raft — composed of pumice rock — covered some 400 square kilometers (154 square miles) of the south-west Pacific Ocean, but months later satellites recorded it dispersing over an area <twice the size of New Zealand itself>." (The preceding link contains some interesting reports from this expedition.)

"The picture below is: High-resolution seafloor topography of the Havre caldera (Rebecca Carey, University of Tasmania/Adam Soule, WHOI)"

Under the surface, the sheer scale of the rocky aftermath took scientists aback when they inspected the site in 2015, at depths as low as 1,220 meters (4,000 feet).

Quoting further: "The caldera, which spans nearly 4.5 kilometers (about 3 miles), discharged lava from some 14 vents in a "massive rupture of the volcanic edifice," producing not just pumice rock, but ash, lava domes, and seafloor lava flows. It may have been (thankfully) buried under an ocean of water, but for a sense of scale, think roughly 1.5 times larger than the 1980 eruption of Mount St. Helens – or 10 times the size of the 2010 Eyjafjallajökull eruption in Iceland". The above photo shows significant heat sources persisting into 2015.

Example#2: From <https://volcanohotspot.wordpress.com/

2017/06/16/1999-2001-eruption-sequence-in-the-gakkel-ridge/ >

Quoting from the above source "The Gakkel Ridge is an extension of the Mid-Atlantic Ridge (MAR) across the Arctic Ocean between Greenland and Siberia. The ridge is located approximately 3800 m (11,000 to 12,000) feet below the arctic ice pack north of Svalbard and Franz-Josef Land".

Quoting further: "The Ridge separates the North American and Eurasian Plates. It is the slowest spreading center known. Prior to 1999 it was thought to be non-volcanic primarily due to the slow speed of the spreading and the lack of seismic events".

The above source reports that there have been 3 western expeditions to the area, first in 1998-1999; second in 2001, and the third in 2007. Again quoting; "The most recent found large surprises in terms of volcanic activity, hydrothermal activity and life around the hydrothermal vents." See: <https://volcano.si.edu/volcano.cfm?vn=377020> for the Smithsonian Institutes report on this expedition (including maps of the area and descriptions of what was found).

The above referenced source (https://volcanohotspot.wordpress) further reports "a massive melting of the Arctic Ocean ice pack following the discovery of a plume of warmer, chemically different water, following the 1999-2001 series of eruptions. The size and longevity of the plume was interesting and unexpected".

Authors note: NASA summarily dismissed the significance of these eruptions because, the heat released was inconsequential "when the total volume of the arctic ocean is considered". May I remind NASA, that from the tip of NW Alaska to Svalbard, is approximately

2,200 miles, across an ocean which is, in many places, 8,000 to 12,000+ feet deep. This NASA statement ignores the physics of how hot water moves quickly to the ocean surface, then spreads widely (might it also reflect a little global warming bias sentiment)?

NASA and other global warming advocates have frequently pointed out that the excess heat absorbed by the Arctic Ocean has resulted in measurable water temperature increases to depths as great as 500 feet (more recently to 700 feet). The Arctic Ocean has significant portions of its extent that are deeper than 10,000 feet. NASA does not consider the "total volume of the Arctic Ocean" in these reports. It is either bias or ignorance and these are not ignorant people (read Vaughns definition of "Normalization of Deviance" in Appendix #1 p 260; It applies in this situation)

Many thought that, because of the ocean depth where the volcanos were found by the Smithsonian Institute expedition (See: <https://volcano.si.edu/volcano.cfm?vn=377020>), that the heat would not rise to the surface. The expedition found evidence that at least one "pyroclastic event," or in layman terms, an explosive event occurred leaving volcanic glass shard scattered widely in the area of the two active volcanos.

Authors note: The combination of hot water, sulfur and carbon dioxide from volcanic activity, when heated by liquid magma could, theoretically, create pressures of: water = 3260 psi + carbon dioxide = 1,071 psi, and sulfur 1288 atmospheres = 18,928 psi. According to the universal gas law these pressures are additive. (*sulfur has 3 critical points, Sm = rhombic, Sl = monoclinic, and Sg = liquid S. At 151°C the true triple point, the three states of sulfur were in equilibrium, all existing at the same time, with a potential pressure of 23,259 psi (sulfur = 18,928 psi, plus water 3260 psi, and CO_2 1071 psi).*

When the pressure of the carbon dioxide from the magma was added to the pressures of sulfur, water, carbonate, bicarbonate and dissolved carbon dioxide at temperatures above the critical point of these substances, the total pressure would be well above that required to bring the heated water to the surface (as reported by the Smithsonian Institution). Once the upward welling of the super-heated water begins to move, it would continue until the heat source was exhausted.

Additional Authors Note: I drafted the above approximately two years ago, just before I began to work on the simplified version of this book; "Fatal Flaws: How the Assumptions, Materials, and Methods Invalidate the Theory of Global Warming". In light of this may I remind you of the following that appeared at the end of the introduction to this chapter. (In the late development department): From various news sources: UNDERSEA VOLCANIC ERUPTION SEEN FROM SPACE: An undersea volcano erupted on Jan. 15th (2022), near the Pacific Island of Tonga. On shore, people rushed to higher ground (earthquake activity and other indicators provided advance warnings which were generally heeded). This volcano exploded with enough force to send significant volumes of volcanic debris 60,000 feet into the atmosphere. The pressures sent atmospheric shock waves (sonic booms) that were seen in videos and on barometers as far away as Florida. News headline From: <https://www.newscientist.com/article/2304822-volcano-eruption-in-tonga-was-a-once-in-a-millennium-event/> The underwater Hunga Tonga-Hunga Ha'apai eruption has already triggered a tsunami, a sonic boom and thousands of lightning bolts, "ash" fall, and will lead to acid rain (rainfall is usually acidic {pH less than 7.0} unless there is significant sodium [Na+] in the ash).

Additional Authors note: The sulfur compounds that were thrown into the stratosphere (and lower mesosphere) were sufficient to color rare cloud formations around the Arctic where there are more than twelve million square kilometers of the sky that are at or below -85^{0}C because of the low solar irradiance since 2018. This has caused rather rare cloud formations to rival the "northern lights" in intensity (these clouds of water and sulfur have a different color and luminosity than the aurora).

The heated water during volcanic eruptions or magma flows is heated by a source between 1,000^0 and 2,500^{0}F. As reported above, it can produce explosive events against pressures equivalent to the greatest of ocean depths, if appreciable sulfur is contained in the magma.

Hot water rises vertically through the ocean tending to remain localized until it reaches the surface where it spreads widely on ocean currents, or if it rises under the ice cap, it tends to melt the ice, and delay ice formation the following winter.

The location of the Gakkel Ridge, where the events described above took place, was under the arctic ice cap north of Svalbard Norway. For several years there have been significant areas of open water north and east of Salvbard following the 1998-2007 period. Those open water areas, decreased slightly over time, to the point that the ice-extent as of 12/1/2021 was close to the 1980 to 2010 average. It will require several more years, or decades, before the total ice volume in this area is equal to the pre-1998 level (1/25/2023 Earthquake activity in Jan Meyer and along this ridge has picked up over the last 2 years and we are seeing an increase in the open water extent in this area {see <fatalflaws.net>}).

Given the location and timing of these events, how could the global warming/climate change advocates have missed or ignored this? It was known and reported by several sources (Including the Smithsonian Institute).

The reason; This information provides an alternative explanation for the melting of a significant portion of the polar ice cap that has occurred. This ice melt is an expression of warming of the earth's surface that is not human caused and has nothing to do with atmospheric CO_2 and other minor greenhouse gasses.

Authors Note: If one googles "earthquakes north of Svalbard Norway" you will find maps of recent earthquakes along the mid Atlantic ridge in the area mentioned above. In the last 5 months a total of 10 significant quakes (4.4 - 4.8 magnitude) were recorded along this Gakkel Ridge area. This is 24 years after this area became active, and is a strong indication that there is still significant magma flow or volcanic activity in the area. It is difficult to monitor what is actually happening because the ocean floor is 10,000 to 12,000 feet below the surface and under the ice cap for much of the year.

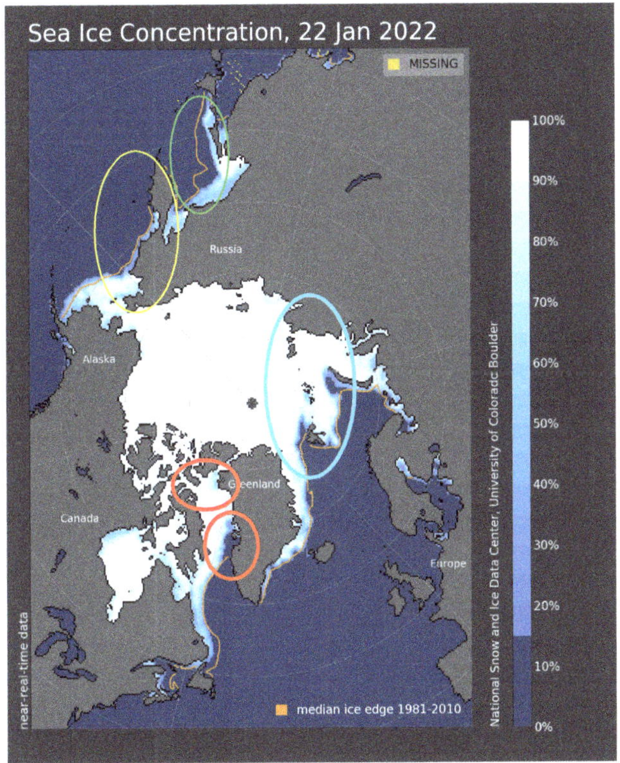

The Arctic Ocean ice cap has been significantly reduced around this area, extending toward the North Pole and Franz Josef Land to the East (manifested by slow consolidation of the ice cap during the early winter).

This melting of the arctic ice cap by volcanic heat is in addition to the 40-50% attributable to "natural variation" reported by <Lawrence Livermore National Laboratory> staff, as discussed in the introduction. It is also, not connected to the direct heating of the Arctic because of degradation of atmospheric ozone as reported by Dr. Polvani who is a Geophysicist at Columbia University, discussed in Chapter 1.

Authors Note: The illustration on p.75, from The National Snow and Ice Center at The University of Colorado, shows the extent of the Arctic ice on 1/22/2022. I overlaid five circles or ovals in areas where there are frequent, shallow earthquakes, and where the ice is slow to develop during the early winter, or may completely disappear at times during the winter when the air temperatures are well below the freezing point of sea water. (2/2/2023 for updates visit <fatalflaws.net>)

The teal colored circle outlines an area around the Kuril and Sakhalin Island group (north of Hokkaido Japan). Within this area small, relatively shallow earthquakes are very common (10 have occurred in the last month {date 2/3/2022}). Magma intrusion (heat source) drives tectonic plate movement and usually is accompanied by small earthquakes of 4.8 or less magnitude. As of Feb 12, 2022 this area contains a highly significant portion of the Arctic that lacks ice cover vs. the 1980 to 2010 average.

The yellow colored circle outlines an area around the Kamchatka region of eastern Russia northward to the Komodore basin. To the south, east and to a lesser extent the west of this peninsula, small, relatively shallow earthquakes are common (10 to 20/month).

The two red circles outline Nuuk (Greenland) and the northern one outlines the Grise Fiord (Canada) and Thule Airbase area of Greenland. Nuuk is an area of thermal activity and the Grise Fiord area looses its ice cover several times each winter even though air temperatures are below -20ºC for 4-5 months 24/7. Ask yourself; Where does the heat come from?

The illustration below, (p. 77) shows the Sea Ice Extent on 1/28/2022 approximately 6 days after the photo on p. 75 was taken. The light blue oval in both photo illustrations reaches from Salvbard Norway (Islands slightly below oval center), Frans Josef Land (Islands near the oval center) to Cape Chelyuskin, Russia (Islands near the upper end of the oval).

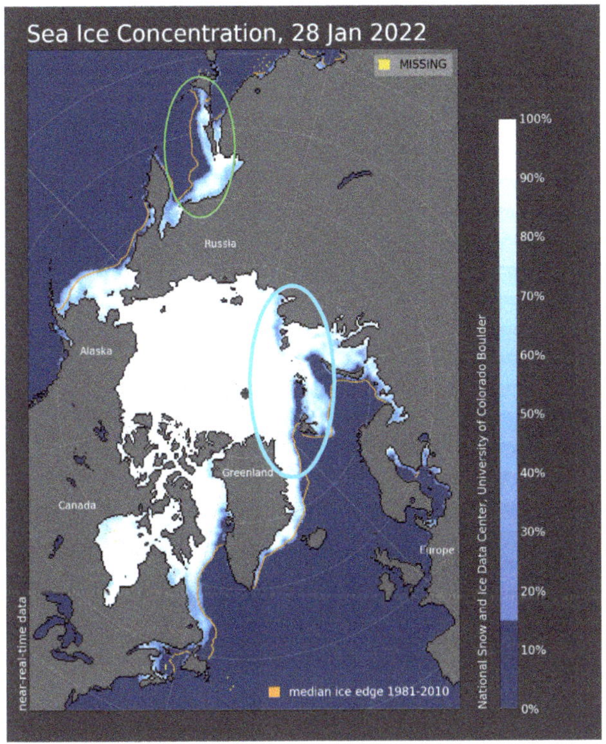

If you compare the two photos you will note that within the light blue oval there has been a significant amount of ice loss in two areas, in six days. The area to the left of Cape Chelyuskin which was at 100% (fully consolidated) has lost nearly all of its ice cover next to these islands and much of the ice cover for a significant distance toward the top of the oval (northeast).

The second area of ice cover loss is just northeast of Salvbard extending to Frans Josef Land (area outlined with light blue oval). This area was approximately 80% consolidated on 1/22/2022 and then was completely open and its width expanded six days later.

During this six day period the air temperature in the vicinity of Cape Chelyuskin varied between 5^{0}F and -7.6^{0}F. The air temperature at Frans Josef Land varied *from -8°F to -28°F during this period. So what is the significance of the two photographs?*

There was a highly significant loss of sea ice coverage in a very short time while the temperatures in the two areas were well below freezing. It should be noted that the "light blue area" is close to the Mid-Atlantic Ridge. The western end of this area, toward the bottom of the photo, is directly over the Gakkel Ridge where earthquakes are frequent and at least two active volcanoes exist 11,000+ feet below the surface of the Arctic Ocean. I will let you consider; Where did all the heat required to melt that much ice in such a short time come from, with air temperatures well below freezing?

The interaction between five environmental forces: 1: The suns temperature, 2: ElNino caused melting, 3: The increased intensity of X-ray and ultraviolet light resulting from ozone depletion, 4: Heat from earth's core, and 5: The greenhouse impact of the minor greenhouse gasses illustrates how multiple forces or factors interact to control the weather and climate in the Arctic and across the world.

Authors note: The impact of the afore mentioned factors or forces is multiplied by the long day length during the "summer months" in the Arctic and Antarctic. Also the lower cloud cover resulting from the abnormal high pressure areas caused by conditions in the tropics for the last 38 years has increased the impacts of these forces. This explains why the polar regions have seen approximately 2.5 times as much warming as the remainder of the earth. The total world wide anomaly, is still only 1.1°C even with the fudged data, or approximately 50% of that with the non-fudged, real numbers. (See Chapters 9 & 10)

<u>This type of multi-variant interactions can't be analyzed or understood by the use of single dimension climate model systems.</u> These models are used almost exclusively in climate science to describe the root cause of earth's warming; With the result that all the warming influence of these 5 forces are attributed to carbon dioxide forcing.

Example #3: Spread across the western side of the Antarctic land mass is one of the largest volcanic areas in the world. Until recently not much was known about this volcanic area but new techniques and technology have allowed scientists to look through the 3,000 to 12,000 feet of ice and find 138 volcanos.

It is not known how many of these may have some degree of activity and for several years, there has been speculation that even one of these volcanos becoming active, could have devastating effects on the Western Antarctic ice sheet.

As it turns out, we have been seeing the effects of that happening during the last several years. All of the impending collapse of this ice sheet, which is one of the three largest in the world, has been attributed to CO_2 forcing of the earth's air temperature. Again we are finding out that this is not true (seems to be a definite trend). Consider the following: From NSF site:<https://www.nsf.gov/news/news_summ.jsp?cntn_id=295861>June 27, 2018

"Tracing a chemical signature of helium in seawater, an international team of scientists funded by the National Science Foundation (NSF) and the United Kingdom's (UK) Natural Environment Research Council (NERC) has discovered a previously unknown volcanic hotspot beneath the massive West Antarctic Ice Sheet (WAIS)". Researchers say, " — The newly discovered heat source could contribute in ways yet unknown to the potential collapse of the ice sheet."

"The scientific consensus is that the rapidly melting Pine Island Glacier, the focal point of the study, would be a significant source of global sea level rise should the melting continue or accelerate. Glaciers such as Pine Island act as plugs that regulate the speed at which the ice sheet flows into the sea. The heat source, Loose and team note, is about half that of the active volcano Grímsvötn, in Iceland."

"While the effects of volcanic heat on the Antarctic ice sheets is an active topic of research, this study provides the first geochemical evidence of a contemporary volcanic heat source, emphasizing the need to detect and understand volcanism, including in models of ice-sheet behavior. The greater understanding of volcanism could alter scientists' perception of the mechanics of ice-sheet loss, including in the areas where the glaciers meet the sea."

"Our finding of a substantial heat source beneath a major WAIS glacier highlights the need to understand subglacial volcanism, its interaction with the marine margins and its potential role in the future stability of the WAIS," (Published in: Nature Communications.)

We have been told the science is settled. We are also told that 100% of the warming of earth's surface is attributable to climate forcing caused by increasing atmospheric CO_2 and other minor greenhouse gasses, again, it appears that is not the case.

This ice melt is another expression of warming of the earth's surface that is not human caused and has nothing to do with atmospheric CO_2. Now do you understand more clearly, my statement that, there is no such thing as settled science?

Authors note: Recently (1/2/2023) there have been a number of news stories about the melting of the Thwaites Glacier in West Antarctic. In 2020 it was discovered that this melting was being driven by warm water circulating underneath this glacier. In late 2021 it was announced that NSF and NERC (the same agencies that sponsored the previous expeditions to study the melting of the Pine Island and Thwaites glaciers) were sponsoring an expedition to study Thwaites.

Typical of the news headlines are: From USA Today: Collapse of Florida-sized glacier may happen soon, raising sea levels and threatening costal cities (Jordan Mendoza, USA TODAY: Jan 27, 2022). From Rolling Stone: 'The fuse has been blown and the doomsday glacier is coming for all of us' (by Jeff Goodall Dec.29, 2021). The Thwaites Glacier is a relatively narrow strip of Glacier that lies to the west of the Pine Island Glacier mentioned above (along the ocean margin).

The amazing thing I have noted in scouring the available announcements of expeditions and the hoopla about rising sea levels, is the absence of any mention of; "Tracing a chemical signature of helium in seawater, an international team of scientists funded by the National Science Foundation (NSF) and the United Kingdom's (UK) Natural Environment Research Council (NERC) has discovered a previously unknown volcanic hotspot beneath the massive West Antarctic Ice Sheet (WAIS)". I would reasonably ask: Why is there no mention of the obvious, i.e., volcanic activity? At this point I think you know the answer.

Example #4: This may be getting old, but least detractors say, it is not happening, or it is not happening often, allow me to give you another example, then I will refrain from more. <https://sealevel.nasa.gov/news/66/beneath-antarctic-ice-lakes-floods-and-flowing-water>

This article concerns a broad swath of the East Antarctic ice sheet, another of the three largest ice sheets that make up the Antarctic ice complex. The first two paragraphs of this article read: "A broad swath of the East Antarctic ice sheet collapsed like an icy puddle on a sunny day, leaving behind a 10-mile-wide crater—evidence of a massive subsurface flood captured by two NASA satellites in 2007 and 2008."

Photo above; "A buried saltwater reservoir could help color "Blood Falls," seeping from the end of East Antarctica's Taylor Glacier." (image credit: Peter Rejcek/ Antarctic Photo Library/NSF)

"A decade of observation has revealed lakes and streams hidden beneath Antarctic ice. It was one of a suite of observations over the past decade that has irreversibly altered scientific understanding of Antarctica's subsurface lakes."

"Instead of static, isolated pockets of water trapped within the ice layers, the new picture—gradually revealed by space and airborne instruments, ground-based GPS and increasingly sophisticated computer models—is one of intricate and dynamic hydrology. Rapidly flowing subglacial streams channel water from one hidden lake to another, sudden drainage causes massive collapse, and the meltwater speeds the slide of mighty glaciers toward the sea."

The above describes a dynamic and changing physical situation that is occurring under a massive ice sheet in the Antarctic, once again proving that the science is not settled as claimed by global warming disciples.

Scientists recently learned that, at the end of the last ice age, with temperatures significantly cooler than they are now, an area of the Antarctic approximately the size of the state of Montana suddenly and completely lost its ice cover.

We are now seeing dynamic changes in the West Antarctic and the East Antarctic. These dynamic changes are fueled by energy (heat) from somewhere. It is not CO_2 climate forcing. This is occurring under thousands of feet of ice, so what is the heat source and how extensive is it? The logical answer is, that with 3,000 to 12,000 feet of ice insulating the Antarctic land surface, the heat must be coming from the earth's core, not CO_2 climate forcing.

Author's Note: A June 28, 2018 report stated that, the coldest temperature ever recorded on earth was eclipsed by a new record of, -144°F (-97.8°C). See:<https://www.forbes.com/sites/trevornace/2018/06/28/its-almost-like-another-planet-coldest-temperature-on-earth-recorded-in-antarctica/?sh=1a566d0b4054> This temperature approaches the coldest theoretically possible temperature for earth and is only 7.3°C above the freezing point of CO_2 {dry ice}.

Another area of wonder I have, is the source of the energy that drives the El Nino cycle. We are told that nutrient dense, warm water rises ("wells up") to the surface along Central and Northern South America, then is blown by equatorial winds westward toward Indonesia. This provides the driving force for this phenomenon.

Authors note: I recently read an article about how the driving force is "cold nutrient rich water rising to the surface," this is in error. How could cold water warm the surface temperature of the seas over vast areas along the equator? ElNino is defined by the rise in surface temperatures of the seas in certain areas along the equator, to 20 degrees north and south latitudes.

Question: Does volcanic and magma heat drive this process? What role does it play? This is one of the most active tectonic plate margins on earth, and there are a large number of active volcanoes along it. <See: <https://www.volcanodiscovery.com/south_america.html> The map shown on this site indicates the status of land based volcanos. The tectonic plate margins closely parallel the shore from Chili to Alaska. This site keeps track of those volcanos (land based) in the process of eruptions.

Yes, we need to find the answers to many questions to fully understand the impacts of heat from the earth's core, on the earth's surface temperature. This is an area which has been hotly debated during the last 25 years. Because we know so little about the under sea and ice cap components of the science involved, we are far from having a complete understanding of these processes and impacts.

We do know that these heat sources are huge and that they contribute to the warming of the earth's surface. The only debatable question is "how much of a contribution do they make?" According to global warming disciples, this warming force does not exist. Rhetorical Question: Are the above circumstances indicative of settled science?

Date: 2/8/2022: What is happening in the Arctic at the present time? The following is my opinion, based on my academic training, professional experience and 40+ years of following arctic temperatures, annual sea ice concentration patterns, and climatic conditions in North America and Canada. This has been done in the context of my need to predict the temperature patterns of the 7 western states because of the importance of temperature patterns on the growth and cropping of fruits, nuts, and grapes .

During the past 50 years, I observed closely the cold, dry conditions during the 1960 - 1979 period, followed by cold, wet conditions of the

1979 to 1990 period. Then the gradual warming that began in the early 1990s which manifested as a reduced number of winter cold spells which were severe enough to damage perennial trees, and also a lesser frequency of crop reducing spring frosts. As LaNina is now becoming dominant, this warming trend is beginning to change. For ongoing updates go to <fatalflaws.net>

Photo: This winter scene, of a ridge (elevation = 9,650') in the Wasatch Mountains of central Utah (USA) is typical of vast areas of the earth's surface which occupy high elevations. These areas are mis-represented in determining the temperature of the earth's surface. (See Chapter 9)

Chapter 5: The Role of Airborne Aerosols in Determining Earth's Surface Temperature

Synopsis: Quoting NASA(<https://earthobservatory.nasa.gov/features/Aerosols>) "Take a deep breath. Even if the air looks clear, it's nearly certain that you'll inhale tens of millions of solid particles and liquid droplets. These ubiquitous specks of matter are known as aerosols, and they can be found in the air over oceans, deserts, mountains, forests, ice, and every ecosystem in between. They drift in Earth's atmosphere from the stratosphere to the surface and range in size from a few nanometers—less than the width of the smallest viruses—to several *tens of micrometers—about the diameter of human hair. Despite their small size, they have major impacts on our climate and our health".*

"Key aerosol groups include sulfates, organic carbon, black carbon, nitrates, mineral dust, and sea salt." "The bulk of aerosols—about 90 percent by mass—have natural origins. Volcanoes, for example, eject huge columns of ash into the air, as well as sulfur dioxide and other gases, yielding sulfates."

An example of natural aerosols are the phenol and terpene materials given off by forest trees which cause the late day "Smokey" appearance in the Smokey Mountains and the blue haze that gives the "Blue Mountains" name to several forest areas in the western United States (they are also the compounds that give the forest that pine scent which is so refreshing).

For example, if one views the mountains between Baker and Pendleton Oregon from I-84 just west of Baker in the late afternoon of a warm sunny day, there will often be a definite blue haze to the air. If there have been thunderstorms in these mountains during the afternoon this haze will be heavy enough to obscure much of the detail of the mountain edges and trees seen from a distance.

Human caused aerosols constitute approximately 10% of the total aerosols world-wide. NASA appears to include some classes of particulates, as having natural origins, that are increasingly (over the last 150 years) becoming more directly related to human activities. Two examples are particulate arising from forest fires 3/4 of which are human caused, and wind-blown dust from deserts and other areas. For these reasons the 10% figure is probably low and increasingly so as time passes.

Aerosols contribute to cooling of the earth's atmosphere as they intercept, scatter and re-radiate back into space some of the suns energy as it moves toward the earth's surface. Global Warming/Climate Change models acknowledge this cooling effect. They do not consider that, approximately 90% of it is not human caused. They also do not recognize that these same aerosols play an important role in the greenhouse effect, thus resulting in a significant over-estimate of the greenhouse impact of CO_2 and other minor greenhouse gasses.

What the doctrine ignores is that these aerosols contribute to the greenhouse effect by intercepting, scattering and absorbing radiant energy given off by the earth's surfaces. This is a warming effect of great significance, most of which is not caused by human activities. The net impact of the warming of earth's surface by aerosols is attributed to CO_2 forcing of the atmospheric temperature. Another fatal fail of the global warming/climate change doctrine.

Chapter 5: To grasp the immensity of this source of climate altering force, consider the rapid expansion of economic activity, including industrial manufacturing, coal fired power plants (with poor pollution controls) and other particulate sources that has taken place in the China-India sub-continent area in the last 35 years. Approximately 38% of the world's population live and work in this area and China is on track to become the world's largest economy in the next 10-15 years.

The satellite photo (p. 88), from NASA; "Shows the pale of human caused particulate and haze over a vast portion of eastern China". The caption reads: "Beijing was completely obscured by air pollution on 10/9/2010".

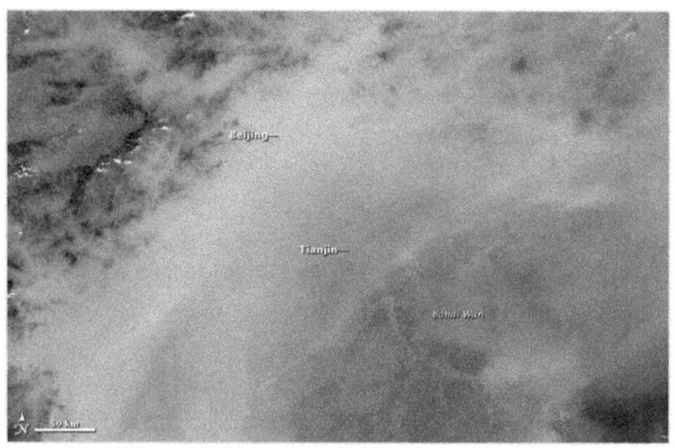

"This natural-color image was acquired by MODIS aboard the Aqua satellite." (Image by Jeff Schmaltz, NASA: MODIS Rapid Response.) "The aerosol layer is composed of complex mixtures of dust blowing from the Thar Desert and pollution from the densely populated plain. In eastern China, fast-growing cities such as Beijing also produce heavy blankets of aerosol."

India, China, Bangladesh and other south eastern Asia nations have been granted exemptions by world bodies (such as the Kyoto and Paris agreements) that allow them to industrialize with little or no requirements to control air pollution (China Produces 29% and India 7% of the worlds air pollution {both countries pollution is still rapidly increasing}).

These exemptions are one of the major reasons that industrial manufacture has tended to migrate away from Europe and the US to these "3rd world countries". For a significant period of time, China, for example, was completing 1 coal fired power plant each week.

Up to the time that the Summer Olympics were held in Beijing, in terrible air quality conditions (an obvious embarrassment to the government), these horrendous sources of pollution were only marginally controlled. There has been some improvement, but insufficient to counter the effects of economic expansion.

Again, quoting from the NASA source referenced above: "However, the most aerosol-laden air in the United States today pales in comparison to Asia. Satellites can detect a visible pall of aerosol clouds over Bangladesh, northern India, and northern Pakistan—an area called the Indo-Gangetic plain, especially during the pre-Monsoon season."

While the average particle, on a weight basis, remains in the atmosphere for a relatively short time (7-10 days). The most active (in terms of greenhouse effect) and representing many times as many particles per pound because of their very small size are the PM_{-10} and $PM_{-2.5.}$

These are also the most dangerous on a health basis, as they are so small they tend to stay in the lungs. These particles may stay suspended in the atmosphere indefinitely unless they are absorbed onto a moist surface such as a rain drop, or the lining of your lungs. They also tend to distribute into the higher layers of the atmosphere where there is little or no water to wash them out.

The net effect of particulate absorption on solar irradiance is a subject area that has not been thoroughly researched. From NASA: "Scientists believe the cooling from sulfates and other reflective aerosols overwhelms the warming effect of black carbon and other absorbing aerosols over the planet. Models estimate that aerosols have had a cooling effect that has counteracted about half of the warming caused by the build-up of greenhouse gases since the 1880s. However, unlike many greenhouse gases, aerosols are not distributed evenly around the planet, so their impacts are most strongly felt on a regional scale".

This cooling effect is included in the estimated attenuation (reduction) of the solar irradiance which reaches the earths outer atmosphere, vs. what is absorbed by the earths surface. NASA and the UN's IPCC estimates that attenuation equals 76% of the total solar radiation that reaches the outer atmosphere. (Water vapor and clouds account for a major portion of this reduction, but aerosols contribute up to 25% of the attenuation.) On average only 24% of the sunlight that reaches the earth's outer atmosphere ever reaches and is absorbed by earth's surfaces.

Authors Note: The cooling effect referred to, occurs as particles absorb or scatter incoming solar radiation, warming the atmosphere at upper and mid-levels, where the energy is readily reflected or radiated back into space, but lowering the total solar irradiance at the earth's surface.

For some reason NASA seems to ignore the other half of the equation, i.e., the effect of aerosols (particulate) on radiant energy released from the earth's surface and the lower layers of the atmosphere as it moves outward. In this case, the particulate in the atmosphere has a definite greenhouse impact. It absorbs out-going radiant energy and then, re-radiates it, with much of it being radiated back toward earth (the greenhouse effect).

The NASA statement above includes: "Models estimate that aerosols have had a cooling effect that has counteracted about half of the warming caused by the build-up of greenhouse gases since the 1880s." This is an additional demonstration of bias as they only consider the cooling effects (which is included in models) as aerosols diminish the solar irradiance at the earth's surface. They neglect their contribution to the greenhouse effect and credit it to carbon dioxide forcing.

The ambivalence toward this topic and the discrepancy is illustrated by the following quote: "Despite considerable advances in recent decades, estimating the direct climate impacts of aerosols remains an immature science. Of the 25 climate models considered by the Fourth Intergovernmental Panel on Climate Change (IPCC), only a handful considered the direct effects of aerosol types other than sulfates".

It is extremely important that one consider the statements in the above paragraphs against the fact that, the energy budgets that are currently in wide usage are based on the solar irradiance that reaches the outer limits of the atmosphere and how much of that, reaches the earth's surface.

These budgets, based on numerous, widespread measurements, of the solar irradiance at the outer limits of the earth's atmosphere and the intensity of solar irradiance at the earth's surface show that only 22% to 24% of the total solar irradiance reaches the earth's surface. Aerosols make a very important contribution to the control of the earth's surface temperature.

They are, along with water vapor and clouds, responsible for limiting the temperature of the earth's surface enough that the surface can support life (this impact is, at most, 10% human caused). Remember the above statement that 90% of the particulate in the atmosphere has natural origins. (Could this oversight be the result of a myopic, biased view of nature?)

The estimate that the earth's surface is currently absorbing "0.85 ± 0.15 W/m^2 more energy than is being emitted into space," incorporates the interception of incoming solar radiation by ozone, sulfates, particulate aerosols, and other classes of particulates and its reflectance, scattering and/or re-radiation back into space. It does not include a separate estimate of particulate interference with the transmission of energy from the earth's surface back into space, which is the particulates contribution to the greenhouse effect.

That impact is attributed to CO_2 and other minor greenhouse gases by default. If you do not consider particulates contribution to the total, but depend on the total, the contribution ends up lumped in with something else, i.e., CO_2. (I realize that I have been redundant, but this is such a typical example of the problem it is worth emphasizing.)

We are left to wonder if the misleading nature of these statements is intentional or merely an oversight. Close examination of the following two statements, contained in the above quoted material, shows that;

Statement 1: "Scientists believe the cooling from sulfates and other reflective aerosols overwhelms the warming effect of black carbon and other absorbing aerosols over the planet". This cooling, mostly occurs in the upper and mid layers of the atmosphere and has been directly included in energy budget models.

In the lower layers of the atmosphere the black carbon and other absorbing aerosols actually have a net warming impact as the result of: First: They absorb solar radiation and by convection warm the air. Second: They are very efficient in absorbing radiant energy given off by Earth's surfaces. Most of this energy stays close to the Earths surface (their contribution to the greenhouse effect). Failure to acknowledge this leads to mis-interpretation of what is, in fact, happening and where it is occurring.

For example the statement above; "Models estimate that aerosols have had a cooling effect that has counteracted about half of the warming caused by the build-up of greenhouse gases since the 1880s". Remember: Ninety percent of this cooling effect is natural (not human caused).

This statement is on its face a farcical attempt to deflect recognization that there is another side of the equation, these aerosols contribution to the greenhouse effect. This tendency to mis-lead is compounded by the 2nd statement.

Statement 2: "However, unlike many greenhouse gases, aerosols are not distributed evenly around the planet, so their impacts are most strongly felt on a regional scale". The earth's climate is not composed of uniform conditions across the entire globe, but it is made up of the sum of varying regional macro-climatic conditions, which are, in turn, made up of the sum of varying micro-climates within regions.

The regional nature of particulate aerosols does not diminish their contribution to the earth's climate. In addition, the PM_{-10} and $PM_{-2.5}$ that finds its way into the upper regions of the atmosphere, tends to stay there, because of the extremely low water content. This is much more uniform and apparently ignored.

This is the same type of critical analysis error that Mr. Gore made in his works when he glibly dismissed the role of water, water vapor and atmospheric ice on the warming of the earth over the past 140 years.

Particulates do play a significant role in the control of the earth's surface temperature through their greenhouse impact. To the extent these impacts are ignored their overall contribution to warming of the surface ends up attributed to the minor greenhouse gases.

Photo (p. 94): Cathedral Valley, at the northern limits of Capitol Reef National Park (USA) is a beautiful spot, but the light scattering effects of aerosols on the view are hard to miss. The snow covered mountains and forests in the background, usually appear somewhat fuzzy, because of the aerosols in the atmosphere. If you are lucky enough to be in this valley immediately after the passage of a cold front moving from the Northwest it is one of the most isolated and beautiful spots anywhere. From the right vantage point you can see five geological structures like the two shown, lined up in an east to west direction. (Caution: before trying to access this valley inquire locally about road conditions. From the South the road disappears in rocky country and from the North you will need to cross several miles of bentonite clay that will mire a 4X4 when wet.) For more on this topic; visit: <fatalflaws.net>

Chapter 6: The Important Roles and Fate of Atmospheric Carbon Dioxide

Synopsis: Carbon Dioxide is one of the most important substances on this planet. Without it life could not exist. Even chemosynthetic bacteria have a physical structure that requires carbon as a basic ingredient.

This gas plays a number of important roles and it is removed from the atmosphere as it preforms these roles, resulting in a dynamic flow of production, utilization, and sequestration. Global warming doctrine focuses on a "bit role" of this important actor, its minor role in the greenhouse effect and entirely ignores some of the "leading roles" played by this gas.

The most important "leading role" atmospheric CO_2 plays, is as a substrate in photosynthesis. In fact, this process, by which plants utilize some of the suns energy to combine CO_2 and water to form sugars, amino acids, and other building blocks of life, uses approximately 706 billion tons of CO_2 from the atmosphere annually. That is approximately 17 times the amount of CO_2 released annually into the atmosphere as a result of human activity.

The efficiency and rate of the photosynthetic process is actually increased by the higher concentration of CO_2 that now exists in the atmosphere. The increased concentration of CO_2 has played a significant role in increasing food and fiber production to support the increasing demands of an expanding population.

In addition, we and the environment have benefited from these higher levels of CO_2 because, increased photosynthesis rates have benefited forests, grasslands, wet lands, and other plant communities that use photosynthesis for the basic building blocks of their growth and development.

Today's agriculture is extremely efficient in establishing plant communities which turn CO_2 from the atmosphere, and water into a vast array of plentiful food, fiber, building materials, paper, and builds soil organic matter, structure, and fertility.

Many different crops are produced using the principles of minimum tillage, cover crops, timely use of small quantities of fertilizer, and other techniques that lead to rapid vegetative growth to establish a healthy leaf surface and then allows that leaf surface to work during as long a growing season as possible.

The elevated carbon dioxide content of the atmosphere is one of the resources that these systems use to produce much more food and fiber per acre than ever before. For example currently 25% of the production of the Washington State Apple industry is grown under certified organic practices. The conventional acreage, utilizes most of the same soil conserving and building practices, which sequester large quantities of CO_2 each year. This sequestered CO_2 is in the form of increased organic matter, and humate which increases soil tilth, water holding capacity, and nutrition elements. (See <fatalflaws.net> for more info.)

The annual net increase in the CO_2 content of the earth's atmosphere is equal to approximately 11.35 billion tons (equals a 2.0 ppm net increase in atmospheric CO_2 per year, which is slightly greater than the 2007 to 2017 yearly average). It is not the 40 billion tons that are released by human activity.

Because CO_2 is not a semi-permeant or permeant addition to the atmosphere, and it goes through both chemical and physical changes in the atmosphere, as discussed in this chapter. It does not qualify as a climate forcing agent, by strict definition.

It should be noted that human activities are not the only factor that causes carbon dioxide to be added to the atmosphere. The increasing temperature of the seas, resulting from increased solar activity and sunlight at the earth's surface, volcanic and magma heat, and El Nino/La Nino *cycles, has caused huge quantities of CO_2 to be released from dissolved forms in the waters of the earth.*

This accounts for a significant portion of the 11.35 billion ton, net annual increase, in atmospheric CO_2 that has occurred in recent years. In addition, as discussed in Chapter 7, changes in the basic doctrines and procedures involved in wildfire control (the let it burn policy) has contributed significantly to the excess carbon dioxide released from sequestered forms during the last 30 years.

At this point, let me remind everyone that, as discussed in the introduction and first four chapters of this book, the role of CO_2 in warming of the earth's surface has been greatly exaggerated. This was done by purposely (or, for the naive, inadvertently) ignoring four highly important forces, and lumping their impacts on warming of the earth, under the CO_2 climate forcing moniker.

For a significant portion of the year, CO_2 is removed from the atmosphere faster than it is added. That may surprise many of you, but for the doubters, you need go no further than the book "An Inconvenient Truth" pages 26 and 27. Mr. Gore shows the CO_2 concentration of the atmosphere for the years 1958 to 2005. The red line goes up and then goes down, once each year.

The ascending portion of this line represents the portion of the year when it is winter in the northern hemisphere and the total, world wide, releases of CO_2 are greater than the total utilization. The descending portion of the line represents the summer period in the northern hemisphere when the utilization of CO_2 world wide, exceeds its release.

The reason that this cycle exists is because the northern hemisphere has a much greater area of land mass than the southern hemisphere. These land masses support more extensive land based plant communities which remove vast quantities of CO_2 from the air through photosynthesis. This is only a part of the story.

Authors Note: From 2013 through 2021 this annual cycle in the global atmospheric carbon dioxide concentration was masked to some extent by the combination of extended severe droughts in various regions, particularly across the western United States, central Canada and Australia. These droughts, interacted with neglected forests resulting in historic numbers of very large forest fires which dumped tens of billions of tons of CO_2 into the atmosphere (see Chapter 7 to learn why and how this happened).

The current rising atmospheric concentration of CO_2 is not the only time that the concentration of this gas has increased dramatically. Pages 46 and 47 of "An Inconvenient Truth" document six dramatic increases in the CO_2 concentration of the atmosphere over the last 600 thousand years. Two important things to remember are:

1. *There were no internal combustion engines, nor other human activities that contributed to the rising CO_2 levels during five of these events, nor during the early portion of the 6th. These events were the result of natural variation, which means their cause(s) were naturally occurring events and processes. These processes are still occurring today.*

2. *Human activity has added an additional 121 ppm to the last rise, but there have been benefits which have accrued from this rise. The increase in the CO_2 concentration in the atmosphere, over the last 140 years, is equal to 0.121 hundredths of 1% of the 'dry' atmosphere. The current total CO_2 concentration of the atmosphere is equal to*

411 ppm, or 0.0411% or, 1/25 of 1% of the atmosphere.

Chapter 6: As stated above, the greatest factor in removing CO_2 from the atmosphere is photosynthesis. All plants, animals, bacteria, fungi, and even viruses are dependent on this process to continue to live. Nuclear energy, wind, hydro, and solar power sources are the only significant energy sources that are not traceable to plant photosynthesis.

As the CO_2 concentration of the atmosphere has increased through the industrial period, the efficiency and rate of the photosynthetic process in most species of plants has also increased. Since 1958 the atmospheric content of CO_2 has increased by 88 parts per million (ppm). This increase has been a significant factor in the increased rate of photosynthesis in most crops, that has helped expand food and fiber production for an expanding human population.

As an example of how increased CO_2 concentrations can increase food production, consider that greenhouse vegetable production facilities will often raise the CO_2 levels to 2,000 ppm (five times the current atmospheric level). At that level, with nutrition, temperature and light conditions optimized it is possible to produce one pound lettuce heads in 21 days, plus the five day period necessary to produce a healthy transplant. At atmospheric concentrations typical of the late 1960s (roughly 340 ppm), it takes 35-40 days from transplant to harvest.

Another highly significant beneficial effect of the faster growth rates of crop plants under the higher CO_2 concentrations in the atmosphere is, that the amount of water used by the plant to produce one pound of end product is significantly decreased. It actually saves water.

Very significant, but virtually ignored is the fact that increased atmospheric CO_2 concentrations increase the rate of photosynthesis in hardwood and softwood tree species. These plant communities sequester a significant percentage of the CO_2 they use in photosynthesis. Softwoods such as popular are very efficient and grow very fast. They sequester a good deal of CO_2 and it is for a fairly long time, particularly if they are used for building materials.

Hardwoods, on the other hand, produce much more complex wood (xylem) structures which are not as efficient in terms of the CO_2 used, but the CO_2 they do sequester endures in the sequestered state for longer periods.

A reasonable estimate of the total use of CO_2 in the photosynthetic process can be found at: <https://earthsky.org/earth/scientists-calculate-a-new-rate-for-global-photosynthesis>. This article, is a report on research done by Scripps Institution of Oceanography, published on September 29, 2011, in the journal, Nature.

It is worth noting that since this research was published the atmospheric CO_2 content has increased by 21 ppm, which has undoubtedly raised the total CO_2 used by the photosynthetic process.

Quoting from the above article: "From their data, they were able to calculate that Earth's vegetation sequesters up to 150 to 175 pentagrams of carbon per year. The following figures are based on the 150 pentagrams of carbon which is 16% lower than the high end of their estimate. In addition the 21 ppm rise in the CO_2 concentration since 2011 would add an additional significant amount to the total.

The 150 pentagram level is equivalent to 175,000,000,000,000 kilograms of carbon. This is a huge number, but notice, it is expressed as kilograms of carbon, so the carbon dioxide used is 3.667 times greater, or 641,725,000,000,000 kg, equal to 641.725 Trillion kilograms. (The 3.667 factor is because almost 73% of the weight of a carbon dioxide molecule is oxygen, which is not included in the 150-175 pentagrams of carbon figure.)

Since most of the figures of total carbon dioxide released into the atmosphere are in terms of metric tons we will stick with the metric system for a moment because it makes the conversion of kg into metric tons easy. A metric ton is 1000 kg (or 2,200 pounds). To convert the 641,725,000,000,000 kg into metric tons we simply drop the last 3 zeros, and presto, we have 641,725,000,000 metric tons. Multiply that by 1.1 and you would have the "English Tons" which equals 705,897,500,000 tons.

This figure is approximately 17.65 times as large as the 40 billion tons of CO_2 released into the atmosphere as the result of human activity each year.

There is disagreement as to how much of the 706 billion tons is sequestered and how much is recycled by the respiration processes of plants and animals and the decomposition of organic matter.

Some of the decomposition of organic matter appears to be included in the 40 billion ton release figure, and should not be double counted; I have made no adjustment for this in any of my figures.

The total CO_2 content of the atmosphere is 3,000,000,000,000,000 kg. The annual uptake and use of carbon dioxide in the photosynthetic process is 641,725,000,000,000 kg. That means that on an annual basis 21.4% of the total atmospheric carbon dioxide is utilized in photosynthesis.

The CO_2 content of the atmosphere is not a static quantity to which an annual increment is added. It is a dynamic concentration which is being added to, utilized and a significant portion sequestered constantly.

The amount of CO_2 above does not include what is taken up by the seas but not utilized in the photosynthetic process by phytoplankton and other water plants that photosynthesize. Nor does it include the CO_2 that is sequestered by the acid-base reactions that form limestone, magnesium nodules, and other forms of complex carbonates.

These methods of CO_2 removal from the atmosphere sequester, for long periods of time, a substantial portion of the 40 billion tons of carbon dioxide that is added to the atmosphere by human activities each year.

As stated above, the fact that the carbon dioxide concentration of the atmosphere is very dynamic means that it is not a permanent or semi-permanent addition. The role of carbon dioxide as a "climate forcing agent" is, by definition, dependent on its semi-permanent or permanent status as an atmospheric component. The same thing applies to Nitric Oxide, Methane, and Ozone. We will discuss this fully later in this chapter.

At this point we should return briefly to the physical properties of water to set the stage for another of its miraculous roles in nature. In this role, it is important in the sequestration of vast quantities of CO_2, by three non-photosynthetic processes.

Water is the most important solvent on earth for many reasons, among them, water constitutes over 60% of the human body and a similar percentage of the weight of most animals and plants. It is the liquid medium in which all life processes take place. (At the USGS site, check out "Water Science School The Water in You: Water and the Human Body".)

Waters capacity to hold a wide range of elements, at various concentrations, in solution is a key element in sustaining life. In addition, over the last 3+ billion years water has played a major role in the physical shaping of the earth on which we live. Water is constantly eroding and dissolving some of the earth's components, then depositing the eroded and dissolved materials in unique and important ways.

Water absorbs large quantities of CO_2 from the atmosphere in a process which occurs 24/7/365. This dissolved CO_2 serves several purposes. The most important role of dissolved CO_2 is as a substrate for photosynthesis in water based plant communities. In fact approximately 50% of all the photosynthesis that occurs on this planet is carried out by phytoplankton and other plants that live in the oceans (this use of CO_2 was included in the total photosynthetic use figures above).

The southern hemisphere has a greater sea surface area than the northern hemisphere but because of a lower level of some important mineral nutrients, particularly iron, the amount of photosynthesis per unit of area is less there.

One of the possible solutions to elevated atmospheric CO_2 concentrations, if they really are a significant problem, would be to fertilize parts of the southern hemispheres oceans by spreading small amounts of chelated iron (and possibly very small amounts of phosphate) across some of the southern hemisphere's sea surface. Some research has already been done on this as a solution, with promising results. Currently additional research is being done, but in an environment in which it is doubtful that a simple, cost effective solution, is desired.

The second major result of CO_2 being dissolved in water is that vast quantities are sequestered by reaction of basic elements with the carbonate and bi-carbonates formed as CO_2 is dissolved. We are all familiar with these products including limestone, coral, sea shells, limestone caves and their various ornaments such as stalagmites and stalactites and other 'hard water' deposits (similar to but much more extensive than, the deposits on your faucets, tub and shower enclosures).

The elements involved in these reactions with carbonates are Calcium, Magnesium, Potassium and Sodium. There has been a good deal of speculation that the amount of material involved in these reactions may deplete the basic elements in the sea and result in a sudden steep increase in the atmospheric CO_2 levels.

This speculation is related to what is termed "acidification of sea water". Quoting from: National Geographic: "Over the past 300 million years, ocean pH has been slightly basic, averaging about 8.2. Today, it is around 8.1, a drop of 0.1 pH units, representing a 25-percent increase in acidity over the past two centuries. (Apr 27, 2017)"

A change in pH from 8.2 to 8.1 is not a change in acidity. It is a change in the base ion [OH^-] so the sea is less basic. The real problem is with the portion of the statement which reads "representing a 25-percent increase in acidity".

The 25-percent figure would be correct if we were talking about a simple solution with no significant buffer(s) in it. <u>That is not sea water.</u> The CO_2, carbonic acid, carbonate and bicarbonate content of the sea is huge (estimated at 38,000 GigaTons; 1 GigaTon = 1 billion tons) or 76,000,000,000,000,000,000 pounds. I would say that is a huge amount, wouldn't you?

When CO_2 is dissolved in water some of it forms carbonic acid. This acid is unstable and dissociates to form bicarbonate and carbonate ions. This gives seawater a great deal of buffering capacity (it resists pH changes in the pH range of the carbonate and bicarbonate ions which is 6.6 to 8.1).

This is what is known chemically, as a dynamic equilibrium reaction. The significance of that is, that as you use up or react some of the carbonate or bicarbonate, the water absorbs more CO_2 from the air, which then reacts to replace the carbonate and bicarbonate that was removed from the water.

Authors Note: It is extremely important to note that the above quote from National Geographic deals with a pH change from 8.2 downward to 8.1. This lowering of the pH of the seas by 0.1 pH units has been outside of the carbonate/bicarbonate buffering range. It will take many times as much "acidification" to lower the pH to 8.0. With the increasing CO_2 concentration in the atmosphere any change will result in highly significant levels of carbon dioxide dissolving into the water; a process that results in long term sequestration. This is one of the reasons I refer to water as a miraculous substance, and its role in modulating the earth's environment as a beautiful thing.

May I illustrate what the above two paragraphs mean? Several times in my professional career I have been asked to help different agricultural operations deal with water quality problems. The World Bank and the Government of Pakistan were among these opportunities in the early to mid 1990s.

Typically, the most difficult situations are where the carbonate and bicarbonate content of the water was high or very high (similar to sea water). In these situations minor nutrient availability is limited. The carbonate actually combines with the nutrient elements to make them unavailable to the plant.

In addition, plant growth is limited by the salt concentration. A solution, in many circumstances, is to react the carbonate and bicarbonate with sulfuric acid which reduces these ions to CO_2 and water, or results in their reaction with basic elements to form insoluble precipitates, thus sequestering the CO_2 (carbonate).

If the carbonate is in the range of 500 - 700 ppm and the bicarbonate is in the range of 150 ppm we can lower the pH to 6.5 (neutralize the carbonate and bicarbonate) by mixing 10 pounds of concentrated sulfuric acid in each 60,000 gallons of water. This is roughly 50 pounds of 36 normal sulfuric acid per 325,000 gallons (1 acre foot). This illustrates how strong the carbonate/ bicarbonate buffer of sea water is. To describe the quantity of sulfuric acid needed to bring the oceans pH down to neutral (pH 7) we would have to define what a #5 number is (do you remember the scale? 1= Tiny, 2= Large, 3= Huge, 4= I can't even say that one.) What should I call a #5 number?

And some of you thought all CO_2 did was give your soda a little fizzy bite.

A third major group of processes that are dependent on CO_2 dissolved in water are the reactions that result in the formation of soil. These are very important reactions and they have been ongoing for 3+ billion years. They have played key rolls in making the earth such a wonderful place to live, work, and play and they sequester a huge amount of CO_2 each year.

If you would indulge me I would like to introduce you to one of the miracles of nature. It is something that has intrigued me, for about 70 years, as I have hiked, rode horses and traveled the back-country (also as I have accomplished research, in the laboratory and field). I had not thought of it as a mechanism that sequesters carbon dioxide until about 25 years ago. It is very important, takes place in billions of places all over the world, is constantly occurring and lasts for ages.

That is the breakdown of rock by dissolved CO_2 (carbonic acid) and organic acids (products of photosynthesis) exuded from plant roots, or released as plant tissues decay after they die. Hopefully you can see the intricate beauty of the process and what it leads to.

As the concentration of CO_2 in the atmosphere rises, more is dissolved into surface waters, and more acid is formed per unit of water (in technical terms the reaction is a dynamic equilibrium reaction). When carbonic acid or organic acids contact rocks or rock fragments, the first thing that happens is that some of the rock is dissolved.

This results in calcium, magnesium or other basic elements, being released. They can react with carbonate or bicarbonate to form precipitates which are quite insoluble. In other words, limestone, coral, stalagmites, stalactites and other related rocks are formed.

That is a long way to describe what happens if you use hard water to make ice cubes. When you place those ice cubes into your glass of water and let them melt, you are left with water which has visible pieces of gray minerals floating in your drink (the layman's term for these is "floaties"). These "floaties" are basic carbonates including minute lime-stones, magnesium carbonate nodules, and other precipitates.

The vastness of the water surface exposed to the air multiplies the amount of CO_2 being absorbed. The higher the atmospheric concentration of CO_2 the faster it is dissolved. Most of the CO_2 remains in a dissolved state (as in Coke, Pepsi , etc.) but a significant amount forms carbonic acid (H_2CO_3). This acid is then free to react with basic elements dissolved in the water or which exist in the rocks and soils the water contacts.

The eroded ledge in the photo above was once CO_2 dissolved in sea water. It then formed complexes with calcium and other bases which have persisted for hundreds of millions of years. It now is at an elevation of over 9,000' in the Wasatch Mountains of Central Utah.

Feldspars and two of the three most common forms of basalt contain high levels of basic elements which are gradually dissolved by carbonic acid. The basic elements can be used in plant nutrition, or they may form precipitates. These precipitates can last for millions of years, resulting in very effective sequestration of huge amounts of CO_2.

I would ask three rhetorical questions; 1. How many times have you seen or been told that photosynthesis removes a large share of the CO_2 released into the air by human activity on an annual basis? 2. How many times have you seen reports or other communications revealing the fact that approximately 70% of the remaining CO_2 released into the environment on an annual basis simply disappears? (Note: there is nothing magical about this disappearance, it involves the processes we are discussing.) 3. Why do you suppose that is the case?

Plants use photosynthesis to form organic acids, some of which are then excreted by the roots or released when plant tissues die. These acids break down rocks and soil particles releasing some of the elements which the plants need for proper nutrition. This occurs basically everywhere plants grow, including in the water.

This process creates free mineral nutrients such as potassium, calcium, magnesium, manganese, iron, copper and zinc which plants can use, or which may react with carbonates to form insoluble precipitates (little rocks) that are different from their parent materials.

The rock breakdown usually begins with the action of lichens (Top photo p.110: pink, blue-green and gray growth on the rock). Then moss (dark plant growth on the rock) gains a foot-hold and when there is enough soil, the grasses, and

broadleaf plants become established as seen here.

These more advanced plants grow larger, have more leaf surface, and more extensive root systems. They increase the speed of the rock breakdown and formation of soil. There is a direct relationship between the amount of leaf surface and the rate of soil formation.

The plant community in the photo below shows another, older community, which includes several pine trees. Eventually, the pines will be growing on a large patch of soil derived totally from the rock through the action of decaying plant parts, carbonic acid, and related plant exudates.

This type of CO_2 sequestration goes on 24/7/365 in billions of places on this great big wonderful world. The perpetual cycle; Plants utilize CO_2 in photosynthesis to make the sugars, amino acids, cellulose and other necessary elements of their lives. Then, plant root systems break down rock by excreting organic acids and respiratory CO_2 which forms carbonic acid.

These acids dissolve portions of the rock, releasing calcium, magnesium and other basic elements that can react with carbonates to form insoluble materials that result in sequestration of the CO_2.

The constant breakdown of rocks by this process is the perpetual source for more basic elements, which then can complex additional CO_2. This, like much of nature, is a beautiful, intricate process; one that is invisible to agenda driven, biased, and myopic science.

The sequence never ends. In the tropical zones of both the northern and southern hemispheres, the temperatures and moisture conditions necessary for plant growth, development and reproduction are at, or near maximum for the entire year.

While there is more total land area in the northern hemisphere, the amount of land surface in the tropical area of the southern hemisphere is significant. Even in arid climates this process takes place. For example miniature barrel cactus and prickly pear can be found growing in soil pockets on and in rocks in eastern Zion Canyon National Park in abundance.

This function of dissolved carbon dioxide in the breakdown of rocks and soil formation is a perpetual process. For this reason the net sequestration of CO_2 by photosynthesis is only part of the story.

Photo: This alpine waterfall among the grass and shrubs is constantly scrubbing carbon dioxide and pollutants from the atmosphere. You will notice that I do not refer to carbon dioxide as a pollutant for good reasons. It is an essential component of the atmosphere which supports the photosynthetic process and performs other essential roles. As carbon dioxide is scrubbed from the air, a portion of it will form carbonic acid which also plays several essential roles in the natural processes that make this earth such a wonderful place to live, work, and play.

At this point the Author would ask for your indulgence as he discusses a concept that illustrates an important characteristic of the global warming/climate change disciples and advocates methods and mindset. Let us begin by discussing something very important from NASA <https://climate.nasa.gov/causes/> that global warming experts (including NASA) get wrong.

Quoting the indicated source: “Certain gases in the atmosphere block heat from escaping. Long-lived gases that remain semi-permanently in the atmosphere and do not respond physically or chemically to changes in temperature are described as "forcing" climate change. Gases, such as water vapor, which respond physically or chemically to changes in temperature are seen as “feedbacks".

This paragraph denotes Water Vapor, liquid and ice as feed-back’s and Carbon Dioxide, (and by inference) Methane, and other minor greenhouse gasses as “forcing” agents.

This paragraph contains a definition of climate forcing agents that the minor greenhouse gasses do not meet. As discussed above, most of the carbon dioxide that is added to the atmosphere by human activities is removed by photosynthesis and other processes on an annual basis. In fact 24% of the atmospheric carbon dioxide is involved in the photosynthetic process each year.

These processes involve both physical and chemical changes, and those changes are temperature sensitive. For example, the solubility of carbon dioxide in water, which is a physical change, and a chemical change, for that portion that is converted into carbonic acid, carbonate and bicarbonate, is temperature dependent.

The second significant thing that is wrong about this paragraph is that water exists in the atmosphere in the vapor, liquid, and solid (ice) physical states. It also exists at much higher concentrations, both on average, and locally, compared to the minor greenhouse gases. Both of these distinctions are of importance.

Quoting the first bullet point of the above referenced article: “Water vapor, The most abundant greenhouse gas, but importantly, it acts as a feedback to the climate. Water vapor increases as the Earth's atmosphere warms, but so does the possibility of clouds and precipitation, making these some of the most important feedback mechanisms to the greenhouse effect.”

This statement contains some truth, water vapor is the most abundant greenhouse gas, but even here a truly myopic view of water and its role is exhibited. The role of water in the atmosphere is multiplied by the fact that it exists not just as vapor, but as liquid and solid. This means that it is not only the most abundant greenhouse gas, but that it is also a much more active substance on a unit weight basis in the greenhouse effect compared to the minor greenhouse gases.

The phrase, "It acts as a feedback to the climate" indicates that this view of waters role in controlling and modulating the climate is viewed from a rather tunnel vision perspective. Water acts as a feedback, but it also is the second most important driver of climatic patterns, after the sun. First, I will describe one of several extremely important feedback functions of water, then I will tell you why water is a driver of short term and long term climatic patterns.

One of the feedback functions of water is what happens when the surface temperature of the oceans increases. As we discussed earlier, in Chapter 2, a vast quantity of water evaporates from earth's surfaces each day. This process requires large amounts of energy, and the evaporation has a tremendous cooling effect on the surface. The heat required to evaporate water, is then carried upward as the water vapor rises to the mid levels of the atmosphere, where it is released as the water condenses.

If the temperature of the earth's surface rises, even slightly, the rate of evaporation of water from the surfaces of the seas, rivers, lakes, streams, and the rate of transpiration of plants, increases. This means that even more heat is moved to the mid levels of the atmosphere where it is readily lost back into space (a cooling process).

Approximately 75% of all the heat loss from the earth's surface occurs in this manner. If the surface warms, the heat loss rate increases, if the surface cools, the heat loss rate slows. This is an extremely important feedback process, probably the most important. The energy involved in the evaporation process moves to the mid levels of the atmosphere as potential chemical energy which can't be intercepted by the minor greenhouse gasses. This process by-passes much of the greenhouse potential of the minor greenhouse gasses.

The dew point and how it acts is also an example of how water, through feedback responses, modulates and regulates temperature changes in earth's weather and climate.

To view water only as a feedback is truly myopic tunnel vision in an astounding way. One of water vapors major roles in the determination of the earth's weather patterns in the short term, and climate patterns in the long term is its role in creating variations in, and controlling the atmospheric pressure. We are all familiar with the terms "High Pressure System" and "Low Pressure System" in describing one of the major weather and climate features. What are they and what causes them? I am really glad you asked. I think I can explain this in "layman terms" with only a little "scientific jargon". I will "keep it simple, but not too simple".

The barometric pressure describes the weight of the atmosphere above a unit of surface area. At sea level, the barometric pressure is sufficient to raise a column of Mercury 29.92 inches. There, I satisfied my urge to be technical, but for those of you who do not know what that means, it is approximately 14.7 pounds per square inch. This pressure varies as the composition of the atmosphere above the surface varies, and the major factor in that variation is the amount of water vapor.

Don't Panic when I mention that, according to the ideal gas law there is only room for a certain number of gas molecules in a given volume. Just like there is a maximum number of people who can ride in my Ford pickup or your mini-van. The number of seat belts determines how many can ride in my pickup, the ideal gas law describes how many molecules can occupy 1 liter of space (at a given temperature and pressure).

Dry air is composed of 78.09% Nitrogen (N_2), 20.95% oxygen (O_2), 0.93% Argon(Ar-40), 0.04% CO_2 with small quantities of other gases. Of course, truly dry air (speaking of the lower 40,000' of the atmosphere) does not exist, there is a variable amount of water vapor, with liquid and solid state water, above the earth's surface.

As mentioned in Chapter 2, the water vapor content of the lower 40,000' of the atmosphere, which is where most of the weather occurs, and is where 99%+ of the greenhouse effect occurs, averages approximately 0.4% (10 times, on a weight basis, the CO_2 concentration). (Stay with me for another paragraph, please.)

Water vapor is much lighter than Nitrogen (N_2), Oxygen (O_2) and carbon dioxide (CO_2). The ratio of the weight of a given number of molecules of Nitrogen, Oxygen, Carbon Dioxide and Water is: Nitrogen (N_2) = 28, Oxygen (O_2) = 32, Carbon Dioxide (CO_2) = 44, and Water (H_2O) = 18. This means that water vapor weighs only 64% as much as nitrogen gas, 56% as much as oxygen gas and 41% as much as CO_2.

This results in a highly significant climatic impact; As the humidity (water content) of the air increases, the weight of the air decreases. This causes the difference between high and low pressure weather systems. It is also why water vapor rises, carrying with it tremendous amounts of energy.

This is the driving force behind winds, and storms such as Tornados, and Hurricanes. This is why my statement that water is second only to the sun as a force that drives weather in the short term and climate in the long term, is absolutely correct!

As an example of how important water vapor is as a primary driver of the climate, please recall the Nov. 5, 2018 publication of the staff of Lawrence Livermore National Laboratory, entitled Models Show Natural Swings in the Earth's Climate Contributed to Arctic Sea Ice Loss: Nov. 5 edition of *Nature Geoscience*. The Arctic high pressure systems that are caused by El Nino have caused 40% to 50% of the Arctic ice melt that has occurred over the last 3.8 decades. This was not and is not the result of a feedback role.

The above source determined that nearly 50% of the Arctic ice melt that has occurred over the last 38 years was driven by the ElNino warming of the ocean surfaces in the tropics. That caused a change in the climate patterns, causing high pressure systems to dominate the atmospheric pressure systems in and around the Arctic.

During this time period the amount of water vapor in the atmosphere has been much greater (approximately 1,100 Trillion pounds). The impact is traceable to atmospheric weather patterns which are the result of water vapor. The impact has lasted for over 3.8 decades which is greater than the length of time required to establish a climatology or climatic trend (that is what is required, though it is arbitrary). This is but one of the climate driving functions attributable to water.

Water plays the role of a primary driver of the climate as well as being involved in extremely important feedback responses. Ignoring the multi-dimensional role of water, results in a grossly distorted, myopic view of its role in determination of climate patterns. If that strikes you as repetitive of other things I have written, I would say; some things warrant repetition.

The other two factors that supports the distinction between "forcing agent" and "feedback" agent are termed here (in this article) as "long-lived" gases that remain semi-permanently in the atmosphere" and second, "do not respond physically or chemically to changes in temperature". We will discuss these two phrases in reverse order, for a purpose.

The phrase; "Do not respond physically or chemically to changes in temperature" is a comparison of how water and these other gases compare in terms of "change of physical state" among other things. You will remember that we discussed how, the fact that water could exist in the atmosphere in either the vapor, liquid, or solid state, increased its greenhouse substance effectiveness by spreading the spectrum of light (energy) it could intercept.

It also means that changes in the atmospheric water content can result in increasing or decreasing solar irradiance levels reaching the earth's surface.

We discussed how waters ability to exist in the atmosphere as clouds and ice also meant that the amount of water in the atmosphere dwarfed the amounts of all the minor greenhouse gases. These other gases are limited in this sense, because they will remain in the gas state within the range of temperatures that will allow life to exist.

My apologies, for getting a little technical at this point. I do so to illustrate how wrong disciples of global warming dogma often are. The above phrase is also incorrect in the assertion that these minor greenhouse gases do not respond physically or chemically to changes in temperature; They do.

For example, the solubility of CO_2, methane and nitrous oxide in water, increases as the temperature of the water (including rain drops which wash them from the atmosphere) decreases.

Authors Note: At https://www.engineeringtoolbox.com : You can find graphic charts that show the effect of temperature on solubility of several of the minor greenhouse gases. You will find that as the temperature falls from 21^{0}C (68^{0}F) to 0^{0}C (32^{0}F) the solubility of CO_2 in water increases by 2.26 times.

Note: That large of an increase in dissolved CO_2 will result in significant increases in the carbonic acid, carbonate and bi-carbonate formed in the water. This is a physical change in solubility and a chemical change in molecular structure, as a response to a change in temperature, which occurs throughout the waters of the earth. The phrase; "Do not respond physically or chemically to changes in temperature" does not describe CO_2, Methane, or Nitric Oxide. How wrong can one be?

The formation of cool or cold raindrops, which then fall through the atmosphere, will remove significant quantities of these gases. Some of these gases are changed chemically as they dissolve. These changes may be hydration (a significant change) or in molecular structure, as carbon dioxide changes into carbonic acid, carbonate and bicarbonate. Other minor greenhouse gasses go through similar changes at a molecular structure level.

Once the raindrops hit the ground or water surface they can and do go through additional significant chemical changes, the rates of which, are temperature dependent. *(For example Nitric Oxide forms HNO_2 {Nitrous acid} which is then changed to NO_3 {Nitrate}, in the soil by action of specific microorganisms. The NO_3 is one of the primary sources of nitrogen for plants.)*

It is surprising to me that anyone can regard atmospheric CO_2, nitrous oxide, or methane as permanent, or semi-permanent atmospheric gases. They are dynamic, ever changing, very small portions of the atmosphere, which are washed from the atmosphere by water, used from the air as CO_2 is used in photosynthesis, or undergo a broad range of acid/base and free radical reactions. For example when nitric oxide dissolves in water it forms nitrous acid {HNO_2}, a highly significant molecular change. Given the definition of atmospheric forcing agents none of these three gases qualify.

The cooler the raindrop, or other water, the greater the absorption of these gases (temperature dependent). The warmer the soil or water surface the more rapidly they undergo chemical changes in the soil or aqueous environment (temperature dependent). Now that wasn't so painful. But it demonstrates how myopic bias can hide the truth.

During periods such as El Nino or the surface warming of the Atlantic Ocean during Hurricane season, there are vast quantities of CO_2 released into the atmosphere because, as the water temperature increases the solubility of this gas decreases. This is a physical/chemical response to temperature and adds significant quantities of CO_2 to the atmosphere.

Processes, such as photosynthesis, soil formation from rock, CO_2 being washed from the atmosphere as it dissolves in precipitation and the formation of carbonic acid, carbonates and bicarbonates, strongly illustrate the untrue nature of the above quoted statement.

These gasses are not permanent or semi-permanent additions to the atmosphere. Let's not forget that it is the relatively short cycle time water spends in the atmosphere between evaporation and precipitation that makes it an efficient driver of the process that circumvents most of the potential greenhouse impact of the minor greenhouse gases.

Rhetorical Questions: How could supposed experts (educated persons) make a series of assumptions that lead them to such magnificent errors? Does bias result in intellectual blindness? Or is bias, intellectual blindness?

I asked for your indulgence while I reviewed the above "matter". There was a reason both to ask for your indulgence (many undoubtedly found it boring), as well as to review the "matter". That reason is; The mindset expressed in the material quoted illustrates the narrow, myopic, view of natural phenomena and processes which are interpreted as simple, single dimensional interactions, or physical reactions, when they are multi-dimensional.

To see the real picture one must take into account all of the multi-dimensional interactions and processes that make nature such a beautiful thing. <u>It is not merely a series of unrelated single dimensional interactions and/or correlations.</u>

That is the problem that resulted in the "Original Fatal Fail" of the global warming theory. It was viewed as a simple correlation between two factors, not a complex set of interactions between at least five simultaneously interacting factors. (In the Introduction and the above, I quoted Dr. Einstein who said: "Make it as simple as possible, but not too simple".) By now, I hope you see the significance of that quote.

The foundational fatal fail of the global warming theory was very simple. The concentration of CO_2 increased as the temperature increased, hence the one was assumed to cause all of the other. As discussed in the earlier chapters of this work, it is not that simple. Other factors have roles in the process and ignoring them and their roles guarantee a massive "Fail" (in current parlance).

As revealed by detailed examination of most of the interacting variables, the role of CO_2 in the small (but significant) temperature increase that has occurred was distorted and over stated. In fact, the impacts of the major forces driving the temperature increase have been mistakenly attributed to CO_2.

To understand how and why the above phenomena recurs over and over (literally we could examine hundreds of examples) I would refer the reader to Appendix #1: "The Normalization of Deviance and Groupthink" that appears on, p 260. Pay particular attention to Vaughn's definition of "Normalization of Deviance".

Authors Note: Appendix #1 will help you understand what has happened, but it will also help you understand what happens next. Consider that there are now tens of millions of religious zealots who are not only used to exercising blind faith in a flawed theory, but who have an ego, economic and usually a political interest in spreading the "gospel".

We hear over and over, what a threat CO_2 accumulation in the atmosphere is to our way of life and to our very existence. We very seldom hear what CO_2 does to benefit us, and even less frequently do we hear that, just as it is constantly added to the atmosphere, it is constantly utilized, and sequestered (removed) from the air that surrounds us.

It is an undisputed fact that carbon dioxide is being added to the atmosphere faster than it is being removed. In 1958 the level of carbon dioxide in the atmosphere was 312 ppm, and by 2018 (a period of 60 years) the level had increased to 411 ppm. This means that on average, over this period, the concentration of carbon dioxide in the atmosphere increased at the rate of 1.4667 ppm per year.

In Chapter #10 we will discuss how methods such as the GISS Surface Temperature Analysis system, temperature spreading, changing physical raw data and homogenization of data have led to an overstatement of the degree of warming that has occurred. In short, CO_2 is not the villain that it has been represented to be.

(In the, be careful of what you believe department.) https://www.climate.gov/news-features/understanding-climate/climate-change-atmospheric-carbon-dioxide. At this site you will find a graph of the carbon dioxide levels for the last 800,000 years. If you look at the right hand side of the chart you see the line going straight up from 300 to 400 ppm. That figure and many more like it are very deceiving, the total average increase is 1.4667 ppm per year for the period of 1958 to 2005, a total of 47 years (according to An Inconvenient Truth pp. 26 - 27).

Summary: *According to Mr. Gore in "An Inconvenient Truth" it is projected that within 45 years of the 2006 publication date, the carbon dioxide level in the atmosphere would be 600 ppm. It is now 16 years into that period (equals 35.5%+) and the carbon dioxide levels have increased from 375 ppm to 412 ppm. That is a rate of 2.31 ppm per year. To fit Mr. Gores prediction the rate would have to average 5.0 ppm per year.*

The fact that the carbon dioxide average increase is greater for the most recent 13 years compared to the 47 year average is an indication that the rate of annual increase is still increasing, but the total is still 4 one hundredths of 1% of the earth's dry atmosphere. Add in the water vapor and the percentage is less than that. Add in the liquid and solid state water and the percentage is reduced further.

The real question is: What does the increase in atmospheric CO_2 concentration mean in terms of its net effect on the earth's surface temperature. *The answer is that the global warming dogma exaggerates the significance of a relatively minor increase to a minor concentration by attributing all the earth's temperature increase to this single causative factor. (See Appendix #2: p. 267 and Evolution of the Anomaly on: <fatalflaws.net>)*

In reality, a number of significant contributing factors are responsible which include:

1. *Changes in the suns radiant intensity and the increasing sunlight which has been reaching the earth's surface since the solar activity increased significantly 78 years ago. (See Chapter 1)*
2. *Changes in the water content of the earth's atmosphere as a result of the increased solar radiation reaching the earth's surface and the ElNino prominence over the last 39 years.*
3. *Changes in the atmospheric barometric pressure patterns because of ElNino/ LaNino cycles which have not only changed the climate of the Arctic, but precipitation patterns across the entire*

Northern hemisphere for the last 39 years. (In the Southern Hemisphere the same thing has happened.)

4. *Changes in the North Atlantic Oscillation which describes changes in the strength of two recurring pressure patterns in the atmosphere over the North Atlantic: a low near Iceland, and a high near the Azores Islands. This oscillation, driven by ocean (water) surface temperature patterns are linked to warm and dry conditions across the U.S. and Northern Europe, and cold, wet conditions across southern Europe. This oscillation goes through cyclic patterns, hence the word oscillation is appropriate.*

5. *Changes in volcanic activity on land, under the seas, and under the Arctic and Antarctic ice caps.*

6. *Increasing particulate and aerosols in the earth's atmosphere and the resultant increase in their contribution to the greenhouse effect.*

7. *Increased evaporation of water from the earth's surfaces and the resultant increase in the by-passing of much of the minor greenhouse gasses potential impact.*

All of the above impacts have been combined with the impact of the minor greenhouse gasses and attributed to the rising atmospheric CO_2. *This is the result of the basic "Fatal Fail" of the global warming/climate change dogma.*

That flaw was and is the assumption that, because rising atmospheric carbon dioxide levels were occurring at the same time that earth's surface temperature was increasing, the entire *increase was attributable to CO_2, a preposterous assumption! I know, I have been repetitious, but, as I said before, some things deserve repetition and this is one of them.*

The photo above contains all the elements of the processes discussed in this chapter.

1. *Green plants use vast quantities of atmospheric carbon dioxide in the photosynthetic process.*
2. *A significant portion of the carbon dioxide removed is represented by the small log in the waterfall which is mostly sequestered carbon dioxide, as is the organic matter in the soils, cracks of the rocks and other locations.*
3. *The mosses and lichens on the rocks are in the process of using carbon dioxide to degrade the rocks and form new soil, at the same time releasing basic ions which can and do react with carbonates to form insoluble precipitates that sequester carbon dioxide for tens of thousands, even millions of years.*

4. *Finally, the water has the ability to take up vast quantities of carbon dioxide from the atmosphere and sequester it as carbonates and bicarbonates for indefinite periods of time.*

If the rising CO_2 *content of the atmosphere is really a problem, there are effective and cost efficient ways to stabilize and reverse the accumulation. The excess* CO_2 *annual increment is equal to approximately 11.35 billion tons. That sounds like a large number, but when compared to the 706 billion tons of carbon dioxide used as a substrate for photosynthesis on an annual basis it is not such a large amount.*

To fully understand the first sentence of the above paragraph read and contemplate chapters 7 and 8.

Chapter 7: Wildfires, Deforestation and Increasing Atmospheric CO_2

Synopsis: In recent years the average size of wildfires in the US, Canada, and several other major "atmospheric CO_2 *contributors" has been increasing. This source has become, since 1990, a highly significant factor in the annual rise of the earth's atmospheric* CO_2 *concentration.*

Politicians, global warming/climate change disciples, and others have made the claim that the size and frequency of wildfires is the result of what they claim is an approximate 1 degree C rise of the earth's surface temperature. This claim is not supported by any factual foundation.

Beginning in the early and mid 1990s the size of the average wildfire in the US began to increase. At that point the global mean temperature anomaly was 0.37⁰C (average of 1995 and 1996). Since that time, the average size of the 61,375 human caused wildfires has increased from 43 to 119 acres. The size of the 9,946 lightning caused fires has increased to an average of 382 acres (annual averages).

During the 10 years, 2009 to 2018 the annual total acres burned by wildfires, managed and prescribed burns, in the USA has been 11,292,643 acres or 17,644 square miles. Of that total, managed and prescribed burns have accounted for 3,319,343 acres or 5,186 square miles per year.

Why have these large fires became more frequent and violent? The blame rests squarely on changes that have occurred in forest management and fire control doctrine over the last 50 years, and particularly since 1995-96. Beginning in the 1970s, there was a steady "environmentalists" increase in the pressure to change long standing forest management and fire control practices.

The use of selective harvest of dead or dying trees in the publicly owned forests through managed tree sales, has been increasingly opposed, to the point that very little area is approved for these practices. An example of this is what happened in the forests of Idaho and Utah during the drought years of 1999 to 2006. During this time period, the drought stressed trees, especially Engelmann Spruce, and Douglas Fir, which use a lot of water and are drought sensitive died across several million acres.

This resulted in tens of millions of dead, standing, mature trees in the midst of young developing forests of more drought tolerant species of conifers. Attempts were made, by forest managers in many areas, to allow the harvest of these trees through timber sell contracts. The 2005 forest management act attempted to make it easier to remove and use these trees. These timber sells were adamantly opposed by environmentalists and the enabling legislation was never implemented.

The result was that those trees, enough to build millions of homes, stood and mostly still stand, slowly rotting or burning. Many areas with this type of forests have experienced significant numbers of large and very large wildfires.

This same type of extreme fuel loads occurred in many forests from California to the Carolines, and from Florida to Alaska. Not all were associated with drought, though many millions of acres were adversely affected by water shortages. Outright neglect, and forced ignoring of fuel load buildup played a major role in many areas. For example, the Happy Camp Fire Complex in California during the 2018 fire season began in, and burned through, an area which had not experienced any major fuel reduction practices in 126 years. That area literally exploded in a firestorm which destroyed most of the city of Paradise CA, and killed over 80 people.

In this chapter I will examine 3 significant wildfires that occurred in the state of Utah during the 2018 fire season. The amount of sequestered CO_2 *that was returned to the atmosphere and the loss of* CO_2 *sequestering capacity over the period of time it will take to re-establish these forests is astounding.*

It is imperative that proactive forest management fire control practices be implemented immediately. We know how to do it (it was done for 70 years prior to the fire control procedure changes of the 1980s and 90's) at lower suppression costs than we currently experience.

If this is not done we will see continued increases in wildfire destruction during dry periods, and the accumulative affects of the loss of forests on reduced CO_2 sequestration will be immense. During the time required to reestablish these forests, decaying organic residues, which represented sequestered CO_2, will continue to be released. The sequestration loss from a forest fire is multi-dimensional. There is the short term release of sequestered CO_2 during the fire (and for years after) and the loss of sequestration which will occur during the reforestation period (25 to 80 years).

<u>Chapter 7:</u> In this chapter we will take a look at what I call, the hidden major contributor to the carbon dioxide concentration of earth's atmosphere. We are told that the increasing level of carbon dioxide in the atmosphere is driving the temperature of the earth's surface upward, in ways and to an extent, that the earth will be uninhabitable within the life span of many of the young people alive today.

The carbon dioxide in the atmosphere is increasing and the rate has been increasing sharply in the last several years largely due to wildfire's increased impacts. During the last three summers the wildfire acreage in California has totaled over six million acres or 9,277 square miles (2019-2021).

Young students from elementary through college are being terrorized with a secular religion's dogma. This doctrine is telling them that if we do not greatly reduce use of fossil fuels within the next 10 to 12 years the earth's inhabitants will be locked into a fatal spiral. Rising temperatures, rising sea levels, ocean acidification, and other calamities will drastically degrade their living standard and for many, if not all, shorten their lives. Rhetorical question: Is there any question that this is contributing to the emotional, psychological, and even suicidal behaviors of our youth, beginning at increasingly younger ages?

All of this, according to the dogma, is solely the result of the use of fossil fuels to heat and air-condition our living environments, provide our transportation, support the current production of consumer products, produce and distribute our food and provide us with electricity to charge our devices, other electronics, and our future electric transportation.

Global Warming advocates claim the increasingly rapid rise in the atmospheric carbon dioxide content of the atmosphere spells the doom of the human race. Further we are told that there is only one way that we have a chance at survival. All, must sacrifice our living standard, and freedom to make many choices that affect our well being, on the Alter of this secular religion.

This religion's prophets and disciples dictate that all of this must happen or we face certain extinction. This ruling class travels widely for conferences, to communicate the dogma, and for recreation while ensconced in prestigious, well paid positions that support their lavish lifestyles. Lifestyles which must and will be maintained for those who are, in their view, the solution to the problem, at the expense of the masses.

The above is a dire, though accurate, analysis of what is happening in our world today. Rhetorical Question 1: Is it true that we are doomed if we don't give up most or all of our fossil fuel use very soon? Rhetorical Question 2: Is it true that there is only one solution? I would submit that the answer to these two questions is an emphatic no!

Even if rising carbon dioxide levels were the problem there are solutions which do not involve drastic changes in the lives of the 7.8 billion people who live on this earth. There are steps which may be taken in a multi-component approach to slow and even reverse the rise in atmospheric carbon dioxide. The following deals with the most important of these components.

As a foundation for this discussion let us begin with some pertinent statistics from the most authoritative source for the United States, the National Interagency Fire Center (<nifc.gov>) under the Fire Information tab — statistics. I realize that this source covers only a portion of the earth's land surfaces (5%) but it represents a large enough contributor to the increasing carbon dioxide levels in the atmosphere to illustrate the magnitude and cause of the contribution from this source.

The Inter Agency Fire Center reports that the annual average number of wildfires for the US from 2009 through 2018 was 66,898. The average number of acres burned by these fires, was 7,973,300 which is equal to 12,458 square miles. In addition to the wildfires, the annual number of acres burned in managed fires (prescribed or managed burns) was equal to 3,319,343 or 5,186 square miles. The annual total of acres burned by wildfires and prescribed burns, has been 11,292,643 or 17,644 square miles during 2009 to 2018. That is a total of 112,926,430 acres or 176,440 square miles in 10 years.

Authors note: Remember the magnitude of this acreage, as we discuss the effects of three wildfires that burned a total of 152,000 acres in the state of Utah in 2018. The 152,000 acres is 0.135% of the total forests lost in this 10 *year time period. The number of pounds of CO_2 that has been, and will be added to the earth's atmosphere because of the US wildfires and controlled burns over the 2009 to 2018 period is equal to 26,314,614,480,000 pounds. Three things about this number. First: it is over 26 Trillion pounds or 13.112 Billion tons. Second: It is even larger than the number of dollars owed by the Federal Government (as of 5/1/2021), and Third: it represents enough CO_2 to raise the earth's atmospheric CO_2 level by 2.43 ppm or 0.243 ppm per year, from 5% of earth's land surface.*

Authors Note: It should be noted that during the last 3 years the prescribed, managed and controlled burn acreage has increased by more than double what they were in the preceding 7 years. This is because wild-land managers are trying to catch up with fuel loads that have built up over the last 3 to 5 decades due to unreasonable federal and state policies (Managed burns are those fires that are allowed to burn with minimal control efforts. This is what I refer to as the "let it burn policy"). The "poster child" for these unreasonable policies is the state of California where severe restrictions have made proactive fire prevention practices nearly impossible, resulting in almost two million acres lost to wildfires during each of the last four years (2018-2021).

In addition to the above statistics it is important to note that for the 10 years from 1983 through 1992 the average wildfire involved 42.89 acres. For the last 10 years (2009 to 2018) that statistic has climbed to 119.2 acres. This is the direct result of changes in wildfire control policy. If we are serious about limiting the carbon dioxide concentration of earth's atmosphere we must return to the fire control policies and proactive prevention practices of the past. These policies limited the average area involved in each wildfire to less than 36% of what has been experienced during the last 10 years.

Authors note: I recognize that there is some discrepancy in the fire numbers I have cited, because it appears that not all naturally caused wildfires are included in the number of naturally caused fires. It is possible the fires that occur in designated roadless wilderness areas (Under International policies) are not included.

In the "be careful of what you believe department". Consider the following from Wikipedia: "In 1944, the US Forest Service developed an ad campaign to help educate the public that all fires were detrimental, using a cartoon black bear named Smokey Bear. This iconic firefighting bear can still be seen on posters with the catchphrase "Only you can prevent forest fires". Early posters of Smokey Bear misled the public into believing that western wildfires were predominantly human-caused. In Yellowstone, human-caused fires average between 6 and 10 annually, while 35 wildfires are ignited by lightning.

The last two sentences of the above paragraph illustrate the importance of caution when reading statistical information from many sources. The Wikipedia article quoted above, bases their statement that: "Early posters of Smokey Bear misled the public into believing that western wildfires were predominantly human-caused," on an extremely limited sub-set when they limit their statistical analysis to fires in Yellowstone National Park. This area represents a very small portion of the woodlands within the US and the subject statement is directly contradicted by the facts from the most authoritative sources (<nifc.gov> cited, NFS 75%, Interior Dept. 90%).

The top photo on p. 135 (Taken Sept, 2009), is of an area in Crandall Canyon, UT, USA which burned in 1934. The forest had re-established over the intervening 84 years.

In late May and June of 2018 a fire, (Trail Mountain/East Mountain Fire) once again burned the area. Note the area covered by the beautiful aspen trees surrounded by pines. Underneath the aspen trees, there was a full new forest of pine trees which would have replaced the aspens over the next several decades.

The photo below shows the same area, after the 2018, Trail Mountain/East Mountain Fire once again burned the area.

I attempted to get the exact same view for both of these photos, but the unstable mountain at my back made it too dangerous to climb to the original position. By Oct. 2019 (the date of the lower photo on p.135), there were aspen trees and brush beginning to grow in many places.

I grew up near this area and was in this canyon several times each year hunting (deer and elk) or hiking and climbing the mountains. These activities began approximately 20 years after the 1934 fire (1954) and by that time approximately 75% of the entire north slope was covered by aspen trees 10' to 20' tall.

With these photos in mind let us examine 3 other wildfires which occurred in the state of Utah during the 2018 fire season.

The Pole Creek and Bald Mountain fires were started by lightning in the lower elevations on the eastern side of Mt. Nebo (UT, USA) in early August 2018. The winter and previous summer had been one of the driest since the late 1950s. The fuel load in these mountains was known to be very dry and heavy.

From my home, I could not see where the fires started because they were on the opposite side of Bald Mountain. For 5 days we watched, as each afternoon we could see a small tendril of smoke rising above the mountain from both areas.

On Saturday morning I asked my wife (only half in jest) if she thought I should take my shovel, fire extinguisher and bucket of water and drive over and put out those fires. She reminded me that the forest service had to know about them and based on their lack of action were apparently not concerned.

On Sunday the forecast south west winds picked up and both of these adjacent fires blew up. The Bald Mountain Fire, which started in a wilderness area, consumed 18,620 acres of roadless wilderness.

The Pole Canyon Fire eventually consumed 102,190 acres and threatened three communities on the West side of the mountains. These two fires required over $30 million in containment costs.

The Coal Hollow Fire, was started by lightning about the same time (this fire was approximately 20 miles north and east of the Pole Canyon Fire which eventually burned along the West side of Coal Hollow). The Coal Hollow Fire was slower to develop and eventually consumed 31,661 acres.

All three of these fires would have been easily controlled under the protocols and control doctrine in place in the 1970s and 80s. Minimal acres of forest would have been lost and the fire would not have been a threat to anyone.

In addition, for six weeks, approximately 2.5 million residents along the Wasatch Front area and much of eastern and central Utah suffered through terrible air quality. The total acreage destroyed by these three fires was 152,471 acres, which equals 237 square miles.

The following facts are important when one attempts to determine the impact of wildfires on atmospheric carbon dioxide content. I will present here a summary of some extremely important points and concepts.

1. When organic matter burns the addition of oxygen results in 1.35 pounds of CO_2 for each 1 pound of material burned.

2. When a mature forest burns approximately 27,000 pounds of CO_2 per acre, or 17.28 million pounds of CO_2 per square mile is released by the time the flames are extinguished. Typically

this is only a fraction of the total tree structure as the trunk and large limbs usually do not burn.

3. In the 3 fires we are discussing, between 45% and 50% of the total area was occupied by mature conifer forest. The fires in these areas resulted in the immediate release of 1.86 billion pounds of CO_2.

3. The remaining area was covered by pin oak, gamble oak, Utah Juniper (cedar trees), and miscellaneous brush and grass. A total of 720 million pounds of CO_2 was released from these areas immediately.

4. This means that, by the time these fires were extinguished, a total of 2.581 billion pounds (1,290,000 tons) of CO_2 was released.

Author's Note: Most of the fuel that fed these fires consisted of complex, vs simple carbohydrates. When complex carbohydrates such as cellulose, and lignin are involved the weight multiplier would be significantly higher than 1.35 because the amount of oxygen required to 'burn' these chemical structures is higher than simple carbohydrates. This fact makes the above figures conservative.

Once the fire is out, the release of CO_2 does not stop. The main trunk of the trees, and larger limbs are left unburnt in most wildfires. The above ground woody material will decompose gradually, releasing the remainder of the CO_2 over a period of years. This increment will be equal to three times the initial release due to the actual burning of the bark, needles, smaller limbs and organic residues on the forest floor. The above ratio of what is released immediately and what will be released through decay processes varies widely depending on the moisture status of the material burned and its density.

At three times the increment of the initial release, this would be equal to 7,743,570,000 pounds of

CO_2, bringing the total release to 10,324,760,000 pounds or 5,162,380 tons. This release would take place over a number of years which has an important impact which is seldom considered.

If the number of acres destroyed by wildfires is higher than the long term average (the current situation) the total release from this factor will increase year to year for at least 10 to 15 years, then it will level off if the area burned each year remains nearly constant.

The above concept is particularly important because according to the Interagency Fire Center's statistics the mean annual acres destroyed by wildfires during the years 2009 to 2018 was 7,973,000 acres. An additional 3,319,343 acres were involved in prescribed and managed burns each year during this period. That makes the total annual mean acres burned by wildfires and prescribed burns 11,292,343 acres (17,644 square miles). This compares to less than 7,639 square miles as the mean annual area involved in wildfires and prescribed burns in the 1983 to 1992 period. That is a 2.31 fold increase in acres burned annually during the last 38 years.

But wait. The picture is not complete yet, there is a huge additional impact on the atmospheric carbon dioxide level. This is the loss of sequestration of carbon dioxide by the redeveloping vegetation for the next 20 to 50 years until a full canopy of CO_2 sequestering plants is reestablished.

If these fires had been controlled when they were very small the mature plant canopies would have continued to convert approximately 30 pounds of CO_2 per mature tree each year into wood and forest residues that would persist for many years.

This carbon dioxide would have been effectively sequestered.

At 620 trees per acre (70 square feet/tree) this decrease would amount to 18,600 pounds per acre, or 1,272,240,000 pounds of CO_2 not sequestered every year (in the 50% of the area covered by mature forests prior to the fires).

In addition, the 50% of the area covered by pin oak, gamble oak, and other species of brush and grass would have sequestered approximately 50% as much CO_2 per acre as the forest areas. That would mean that each year these areas would have sequestered 777,480,000 pounds of CO_2, making the annual total loss of sequestered carbon dioxide 2,049,720,000 pounds for the 152,000 acres destroyed by these fires.

Over the 20 to 25 years that it will take to restore a full CO_2 sequestering plant canopy over these areas, the annual photosynthetic use and sequestration of carbon dioxide will be less than 40% of what the mature canopies were sequestering before the fires. I initially assumed that the rate would be approximately 50%, then realized that:

1. The species which will constitute the plant community during the early years will be annuals and short lived brush and aspen trees growing from surviving root systems under the soil. These species are not as efficient in long term sequestration of carbon dioxide as are evergreen forest species, oak and juniper (they do not live as long).
2. As the evergreen species begin to grow they will constitute a very small percentage of the total plant canopy for the first 10 to 12 years, then will grow rapidly in terms of canopy size, becoming the dominant species over the 2nd or 3rd decades following the fires.

Based on the above, the 2,049,720,000 pound figure will decrease over time, resulting in an average annual loss of approximately 60% in terms of long term sequestration of CO_2. Using the 20 year figure between the fires and the reestablishment of a mature canopy, the total lost sequestration of CO_2 from this factor, would be equal to 24,596,640,000 pounds (2,049,720,000 lbs/yr X 0.6 X 20 years). This loss would be additive to the 10,824,760,000 pounds lost to the actual fire and decomposition of the residues.

The total lost, long term sequestration of CO_2, resulting from these three fires would be equal to over 35 billion pounds (35,421,400,000 pounds), or 17,710,700 tons.

This analysis does not deal with the CO_2 sequestered by the root systems. In the tree and brush species involved in most wild-lands in the temperate and arctic areas the amount of CO_2 sequestered in the root systems is equal to a major portion of the above ground plant material (70-80%). I have not included this plant material in this analysis because it is difficult to estimate and most of it will gradually be converted into free carbonate and bicarbonate.

This is because much of the CO_2 released into the soil by respiration, or as the result of microbial breakdown will cycle into carbonic acid. Part of the carbonic acid changes to carbonate and bicarbonate, as it breaks down rock and soil particles. This forms insoluble precipitates, soil particles and free mineral nutrients as discussed in the previous chapter. This “soil formation” process takes place 24/7/365 and is actually taking place in living as well as dead forests. The difference is, in a burned forest there is a guaranteed decrease in the future where, in well managed, living forests it can continue uninterrupted, at a maximum rate for hundreds of years.

The three fires we analyzed above covered 152,000 acres which represents 0.013% of the average annual acres burned over the last three years. The total CO_2 released by these fires will be 17.7 million tons.

As discussed earlier in this chapter, the total CO_2 utilized in photosynthesis annually is 706 billion tons and the elevated atmospheric CO_2 content increases this number because the photosynthetic mechanism thrives on higher CO_2 levels.

That can be seen by examining the atmospheric carbon dioxide levels during the 2019 summer in the northern hemisphere. On May 15, 2019, the Mauna Loa Observatory found the atmospheric carbon dioxide level had reached an annual peak of 415.7 parts per million.

When one considers the total CO_2 released into the atmosphere by wildfires, it is not surprising that the May 2019 peak CO_2 concentration of earth's atmosphere reached 415.7 parts per million following a disastrous 2018 fire season. During the summer months, the photosynthetic use of CO_2 from the atmosphere along with the other processes that remove this gas resulted in a decrease in the concentration of CO_2.

The July value was 411.77 parts per million and by September 3, the level had dropped to 408.63 parts per million. The level remained constant through November 15, in spite of the disastrous 2019 fall wildfire season in California and the beginning portion of the Australian fire season (another disaster). These areas are the leaders in widespread adoption of the policies of the International Union for Conservation of Nature.

This illustrates how fast the photosynthetic process, coupled with other sequestration processes, removes CO_2 from the atmosphere. In the above five month period the atmospheric CO_2 level was reduced by 7.07 parts per million. This equals a net 40.1 billion tons of carbon dioxide removed from the atmosphere in a 5 month period from May 15 to October 15, 2019.

Author's Note: Since the early 1990s the pattern for the fire control policy changes has been the 32 million acres of roadless forest lands (within the US and much more world-wide) that are regulated by the IUCN (International Union for Conservation of Nature) regulations.

These regulations played a major role in the 1988 Yellowstone Fire disaster which consumed over 800,000 acres (1250 sq. mi) of forest in and around Yellowstone National Park. They were also a significant contributor to the Bald Mountain/Pole Canyon/Coal Hollow fire complex analyzed herein. These policies explain why we observed the smoke of two of the three small, lightning caused fires, for at least five days, while fire control measures were not used, then it became an uncontrollable wild fire.

A significant portion of the wildfire problems encountered in the last 30 years are traceable to the concepts of the IUCN philosophy being incorporated into the general wildfire control doctrine and procedures. As the principles of these regulations have been adopted over vast areas of the planet, the annual loss of forests to wildfires has increased dramatically (by 283% in the US).

The figures in the paragraphs above illustrate several important facts and also suggest a solution to this problem (to the extent that it is a problem). Those facts are:

A. After the wildfire control procedures where changed between the late 1980s and 1996, the average size of wildfires in the US has increased from 42 acres to 119 acres (a 283% increase), and the numbers of wildfires that burn over 100,000 acres have also increased significantly.

B. The average annual total acres consumed by wildfires and prescribed or managed burns during the last three years (2019-2021) has been 13,703,096 acres or 21,411 square miles. The three wild fires we analyzed above will have a net result of 17.7 million tons of CO_2 released from long term sequestration and lost sequestration over the next 20 years as the plant communities are reestablished.

C. The amazing thing about the facts presented in A, and, B above is that the CO_2 concentration of the atmosphere has not risen even faster. The reason it has not, is the result of the amazing dynamics which buffer the atmosphere by:

1. As atmospheric CO_2 levels increase, the rate of photosynthesis world wide, increases dramatically.

2. As atmospheric CO_2 levels increase the rate and amount of CO_2 which is dissolved into the worlds seas, lakes, streams and other waters increases. (this is currently estimated to be 38,000 gigaton; 38,000,000,000,000,000 metric tonnes {1gigaton= 1 billion metric tons}).

3. As the level of dissolved CO_2 in the earth's waters increases, the formation of carbonic acid increases which in turn increases the levels of carbonate and bicarbonate.

4. The increased levels of carbonic acid, carbonate, and bicarbonate then results in dramatic increases in the rate of very long term sequestration of CO_2.

D. During the 5.5 months (from 5/15 to 10/23/2019) the level of CO_2 in the earth's atmosphere dropped by 7.07 parts per million. This equals 40.1 billion tons (equal to 100% of the annual human caused CO_2 additions to the atmosphere). That is a rate of 1.29 parts per million per month.

Author's Note:The figures in the above paragraph were derived by beginning with the total weight of the earth's atmosphere which is equal to $1.135X10^{19}$ pounds. One part per million is equal to $1.135X10^{13}$ pounds, or 11,350,000,000,000 pounds, which, when divided by 2000 pounds per ton equals 5.675 billion tons. (1 ppm = 5.675 billion tons, 2 ppm = 11.35 billion tons, 7.07 ppm = 40.1 billion tons).

Before 2014 the average annual increase in atmospheric CO_2 concentration was at, or below two ppm which equals a net increase of 11.35 billion tons or slightly more than 28% of the annual human activity related additions. Remember this figure, it is a major factor in solving the problem. The size of the problem is not the 40 billion tons (has increased to 47 billion tons for 2020) that is frequently indicated.

In simple terms, the above discussion shows that wildfires and controlled burns release substantial amounts of CO_2 into the earth's atmosphere. In the last 10 years the annual total releases from this source has increased sharply for definable reasons. The opportunity presented by these facts is, simply stated;

If we truly wish to stabilize the atmosphere's CO_2 concentration the first step would be to do what is necessary to return the annual levels of wild-land fires to something close to the 1983 -1992 levels.To accomplish that goal we would need to:

1. Return to a policy of detecting and extinguishing all fires as quickly as possible (world wide).

2. Institute land management practices aimed at proactively preventing fires, and limiting the size of fires which do get started. This should include extensive harvesting of forest products including dead or dying trees, wood chips and other products that could replace other building materials and energy sources.

3. Revise the regulatory environment to incentivize and require adoption of realistic land management practices on both private and public lands to proactively lower fuel loads to reasonable levels.

If we could decrease the average size of fires that do get started, to a level similar to the 1983 levels we would reduce the number of acres burned by a factor of 2.83 which would take the average annual acres burned down from 13,703,096 (2016 - 2018 average) to 4,842,083 acres. This figure would include approximately one million acres of managed burns which would lower the long term threat considerably. This would, by itself, go a long way to decreasing the annual rise in the atmospheric level of CO_2 to under two ppm

Photo (p. 146): This garden is located at the foot of Black Mesa, south of Kayenta Arizona. The corn, potatoes and squash plants in the background are being raised with modern drip irrigation and "naturally sourced" fertilizer (sheep manure incorporated into the soil the previous fall). In these systems 1,500 and 2,000 pounds of produce can be produced in a 30'x30' garden in what is generally considered a rather hostile environment. Corn, beans and squash are particularly adapted to utilize the higher CO_2 levels in the atmosphere.

Chapter 8: An Affordable Approach to Stabilizing or Reducing The Atmospheric Carbon Dioxide Concentration

Synopsis: The facts presented in Chapter 7 suggest a rather obvious series of actions that could be taken to stabilize or even reduce the atmospheric carbon dioxide concentration. In this chapter I will lay out the general principles, but the detailed methods and mechanics will be published elsewhere (On the website <fatalflaws.net> during 2023). Those actions are:

First: There is a dire need to reduce the numbers and average size of wildfires, not only in the US but world wide. Also fire use in agriculture and other land clearing practices needs to be limited. When an acre of forest is destroyed it needs to be replaced.

Second: The conversion of agricultural lands and Forests into roads, parking lots and other uses which remove the plant cover needs to be limited; or fully compensated for, by replanting forest lands that will replace the sequestration lost in the land use conversion.

Third: A serious worldwide effort at reforestation needs to be undertaken and financially supported by public funds. A system of private investment funding and/or tax incentives could do a lot to replace the nearly 1 billion hectares (2.47 billion acres) of forest lands lost since 1920.

Authors Note: The first three items mentioned above can be put into prospective when one considers the following. If the forests lost since 1920 were replaced this would be the effect: Multiply 2.47 billion acres by 400 trees per acre (100 sq. ft./tree) and you would have 988 billion additional trees. At canopy maturity each of those trees will sequester 30 lbs of CO_2/year. That is a total of 29.640 Trillion additional pounds of CO_2 sequestered each year. The above mentioned 29.640 Trillion pounds of CO_2 is equal to 14.820 billion tons of CO_2. That is 131% of the annual net increase of CO_2 in the atmosphere. This is attainable if wildfires were controlled similarly to the 1983-1992 time period and a meaningful reforestation program is begun.

Fourth: The exemptions given to countries in the India, China, South East Asia area for the last 30 years that allowed them to industrialize with little or no attention to emissions control need to be lifted. They should be subject to the same international pressures to reduce emissions of CO_2 and particulate as other industrialized countries.

Fifth: Tropical agricultural methods need to be altered, to increase the planting of long lived trees for selective, rotational harvest, and to create a more favorable environment for diverse food production. This can be accomplished as part of reforestation projects, making the positive impact on the financial status of participating individuals and communities even greater.

Authors Note: For a number of reasons a major portion of a reforestation program to limit or reverse the rising CO_2 concentration in the atmosphere should be centered in the tropics. These reasons and methods will be part of a future work to be published in 2023.

Sixth: Fertilization of selected areas of the southern hemisphere seas need to be researched, funded, and incentivized to make *those vast areas more productive in terms of CO_2 sequestration, fish and other seafood harvests. Doing this could have major impacts on the earth's atmospheric carbon dioxide content because of the increased photosynthesis this fertilization would stimulate.*

Chapter 8: We have been told that the science is settled; CO_2 is responsible for 100% of the warming of the earth's surface, and that there is only one way to deal with the rising CO_2. Again, as we have observed previously in this work, that is not the case. As additional evidence, a quote from the 2018 FAO, The State of the World's Forests report of the Food and Agriculture Organization of the United Nations.

"As underscored at the Paris Climate Agreement in 2015, forests and trees play a crucial role in determining the accumulation of greenhouse gases in the atmosphere. Acting as carbon sinks, they absorb the equivalent of roughly 2 billion tonnes (metric) of carbon dioxide each year. However, deforestation is the second-leading cause of climate change after burning fossil fuels and accounts for nearly 20 percent of all greenhouse gas emissions — more than the world's entire transport sector."

Authors Note: The above paragraph admits "forests and trees play a crucial role in determining the accumulation of greenhouse gases in the atmosphere." Further it states: "deforestation is the second-leading cause of climate change after burning fossil fuels and accounts for nearly 20 percent of all greenhouse gas emissions."

There are internal inconsistencies in the above figures. The annual total human caused addition of CO_2 to the atmosphere reportedly has been 40-47 billion metric tons. In the last sentence it is stated that: "deforestation is the second-leading cause of climate change after burning fossil fuels and accounts for nearly 20 percent of all greenhouse gas emissions". Twenty percent of 40 billion metric tonnes of carbon dioxide is equal to 8 billion *metric tonnes. The 8 billion metric tonnes of CO_2 is approximately equal to the amount of CO_2 released by burning and above ground residues left by forest fires (wildfires). The lost sequestration resulting from removal of the photosynthetic capacity of the established plant community is ignored. It is actually the largest portion of the lost CO_2 removal from the atmosphere, that results from deforestation.*

The portion of the subject paragraph which reads: "Acting as carbon sinks, they absorb the equivalent of roughly 2 billion metric tonnes of carbon dioxide each year" is in error. The 2 billion tonnes figure is low by a factor of at least 4X. There are currently 9.9 billion acres of forest covering approximately 31% of the earth's land surface. The 2 billion metric tonnes of sequestration per year would only amount to 444 pounds per acre of sequestration. If the average forest contains 200 trees per acre (212 square feet per tree) the 2 billion metric tonnes would mean that only 2.2 pounds of carbon dioxide would be sequestered per tree each year. That is why I wrote above that the 2 billion metric ton figure above is low by a factor of at least 4X. This error is of extreme importance.

Twenty percent of 40 billion tonnes is equal to 8 billion tonnes as pointed out in the second paragraph above. If forest fires were limited, the net sequestration of carbon dioxide could be raised to 10 billion tonnes (2 billion current net plus up to 8 billion from limiting wildfires). That could reduce the excess carbon dioxide released by 25%. If the 2 billion metric tonnes is replaced by a reasonable figure the net sequestration of carbon dioxide could be raised to as much as 16 billion metric tons. That would require total elimination of wild fires which can't be done, so a reasonable estimate would be between 10 and 12 billion metric tonnes.

Authors note: The percentage of the earth's land area

covered by forests is 31%. Approximately 50% of the total photosynthetic use of carbon dioxide (706 billion tons/2 = 353 billion tons) occurs in land based plant communities. There are currently approximately 9.9 billion acres of forest world wide. One ton per acre of carbon dioxide sequestration would result in 9.9 billion tons of carbon dioxide sequestration annually (One ton per acre/year is a very low estimate for well managed forests grown for building materials and other benefits).

Rhetorical Question: With the above in mind how does anyone justify the changes in forest management which have taken place over the last three decades causing wildfires to become the number one cause of deforestation?

Do you realize that you have just read an admission that at least one-fifth of what is claimed to be CO_2 forcing of the earth's surface is due to deforestation?

Do you realize that this figure only deals with the short term decreases in photosynthetic removal of CO_2 from the atmosphere and does not consider the total effects of lost forest areas as discussed in this work?

Worldwide, global tree cover losses rose 51 percent in 2016 from the previous year to 297,000 square kilometers (114,672 square miles), according to data from the University of Maryland compiled by Global Forest Watch (GFW). This source places the total global tree loss at 114,672 square miles annually (equals 73,390,080 acres). Most of this loss is due to human caused wildfires, with smaller amounts due to agricultural expansion and urbanization.

An aggressive reforestation program should be undertaken. The effort should be world wide, but centered in the tropics for 2 reasons. First: It would result in huge economic benefits for millions of the poorer inhabitants of this planet. Second: It would be the lowest cost approach to stabilizing and reducing the atmospheric CO_2 concentration. Properly planned and executed reforestation efforts would quickly become a profitable, self sustaining, perpetual economic blessing to hundreds of millions while vastly improving the earth's environment.

Consider the following facts.

1. Forests scrub (remove) from the atmosphere many different pollutants and greenhouse gases.
2. Forests assist in mitigation of variations in water supplies and quality.
3. Forests cool the environment, an important quality of life factor in the tropics.
4. Use of the right species in properly designed forests can provide a conducive environment for production of food crops such as peppers, sweet potatoes, several types of "spinach," and a significant number of tropical fruits such as passion fruit, etc. They also produce substantial amounts of building materials.
5. Forests build and restore depleted soils and available mineral nutrients within the soil.
6. Forests yield huge economic benefits for those who live and work in and around them. These facts are particularly true in tropical areas. These economic benefits are well documented in the 2012 addition of the above referenced UN Report.
7. Forests located in tropical regions have a much longer annual growing season than in temperate or polar regions. This results in higher levels of annual CO_2 sequestration per unit of land surface.

The world does not need to rely on reforestation to accomplish 100% of the needed sequestration of CO_2. Research has shown that iron fertilization of selected areas of the southern hemisphere seas could have an immediate beneficial impact by vastly increasing the density of plant life in these oceans.

The difference in total photosynthesis per unit of area, between land and water plant communities is generally associated with plant densities.

Where there are a lot of plants in the water, as in the case of heavy algae populations or seaweed beds, the rate of photosynthesis per unit of sea area can be as high as in land based plant communities.

In most of the worlds oceans the density of plants is lower than it is in stagnant or slow moving waters. Less photosynthesis per unit of surface area is the result.

In sea water, particularly in the southern hemisphere, the concentration of iron is very low and is the limiting factor in plant density. Several times in the past five decades there has been research conducted to see if additions of certain forms of iron to the oceans' water will result in an increase in plant density and photosynthesis. The results of these experiments have been impressive with reasonable cost/benefit ratios.

A program of expanded testing, based on the favorable results of small scale efforts, should be funded. This could make these vast areas more productive in terms of photosynthetic activity, fish and other seafood harvests.

Doing this would have major positive impacts on the earth's atmospheric carbon dioxide content because of the increased photosynthesis and CO_2 sequestration this fertilization would stimulate.

For an illustration of what is possible, consider what happens to the fishing industry off the West coast of Central and South America during the El Nino/La Nino cycle. During El Nino, nutrient rich water, wells up from the ocean depths resulting in excellent conditions for ocean plant communities. The number of harvestable fish increases and people prosper and have better diets.

During La Nino periods, the nutrient dense waters disappear, plant communities do not thrive and fish numbers decrease dramatically. It is the nutrient content of the water that makes the difference and the limiting nutrient is iron. For a few dollars worth of the proper iron containing material per square mile each year this nutrient deficiency could be addressed and several problems would be, to a significant extent solved. Those problems include, atmospheric CO_2 concentrations, poverty, and diet sufficiency.

The nutrient rich water during El Nino is warmed by some heat source. As water warms above 4^0C it expands and becomes less dense. This water then rises to the surface. Rhetorical Question: What is the heat source that warms this water enough for it to rise from the ocean depths? The nutrient content would indicate that it could be magma flows along the tectonic plate edge that runs from the tip of South America along the West coast to Alaska (The ElNino is driven by rising, warm, nutrient rich, water along the Peruvian/Central America coastal area).

The water nutrient profile during ElNino is consistent with a rhyolitic magma, one of the different magma classes which is high in iron content, most of which is not water soluble. This limited nutrient supply is sufficient during the active upwelling of the warmed water, but not enough to raise the nutrient content of the water for the long term.

In Conclusion; The above six components could

serve as the basis of a multi-faceted approach to systematically remove CO_2 from the atmosphere in vast quantities. The net costs of these components would be relatively small in the short term (1st 10 - 20 years) compared to alternative versions of the New Green Deal.

In the long term, implementation of these facets in a significantly financed and technologically supported manner could be very profitable. It could be done in ways that would result in the economic benefits flowing to the most needy, particularly in tropical areas, where deforestation since 1920 has been so devastating.

The EU has made some significant efforts at reforestation. That effort is the reason that their forested areas have increased by 7% in the last 28 years. Today, forests cover approximately 30% of the earth's land surface, or slightly more than 4 billion hectares (9.88 billion acres). We have lost approximately 5.5 billion acres of forests since 1700.

Forests, water and plant communities are treasures!

Chapter 9: Global Warming/Climate Change: How The Numbers Work

How is the Earth's Surface Temperature Determined

Synopsis: *The global warming/climate change theory employees a unique system of numbers (anomalies, designed for use in single dimension climate models). How they are generated, analyzed, and the set of assumptions that underlay the structure, use, and communication of what those number's mean, is rather complex. It is imperative that we understand why the number system exists, why it is used, and what its purported advantages are.*

It is also imperative that we understand what the disadvantages and risks associated with this system are. The errors inherent in the system, and the magnitude of those errors, individually and in total, are extremely important. This impacts the conclusions drawn, based on the system, and how much trust may be placed in those conclusions. We also need to know if there are hidden errors, or errors not admitted, or interjected into the system.

On Feb. 12, 2019 an update of a NASA document entitled: GISS Surface Temperature Analysis was published. This is the latest in a series of updates of a document that has been available for 25+ years. This document attempts to explain why the GISS Surface Temperature Analysis system is necessary, why it is better than the alternatives, and in simplified terms, what it is. It is a series of questions and answers, three of which I will discuss in some detail in this chapter.

As you read these questions and answers ask yourselves; does this document make sense? Are there things in it that are not true as it relates to your life? Is this system better than merely using the raw data, determining the global monthly and yearly average surface temperature, consistent with how we use temperature predictions and information in our daily life, and in research in other fields? In short, is the system justified? As I will discuss, there are errors, pitfalls, and opportunities for mis-use in the system that are hidden, not acknowledged, glossed over, ignored, or unrecognized.

The ultimate fatal fail of the system is the wide array and vast magnitude of the changes that have been, and are being made, to the raw data. Scientifically this practice is indefensible. These changes account for approximately 50% of the alleged increase in the earth's surface temperature.

Realize that this system has its roots in the late 1970s and early 1980s (early space age). That is 40+ years ago. Why is this important? Since that time technology has advanced, and satellite systems have been developed which can measure the surface temperature of the earth with much greater accuracy and precision.

Unfortunately the data from these systems appear to be subject to significant data manipulation to reach a doctrinally friendly result. Rhetorical Question: Why do we need a system that has large admitted error factors, and even larger interjected, and ignored errors, based on early space age concepts, when we have at our disposal, systems such as MODIS, SABER and others? Contemplate this; Is any system immune from misuse or abuse? (See Evolution of the Global Anomaly: on <fatalflaws.net>)

Chapter 9: I feel safe in making the statement: "most people do not understand the system that is used to determine the earth's surface temperature when the warming, or cooling (anomaly) of the earth's surface temperature is discussed". It is a system that is different from the one that you and I rely on if you want to know what the temperature is outside, what the average temperature was for the previous week, month, or year. It is not the system that was and is used and relied on by most people for the above purpose, in Japan, the Philippines, Mexico, Europe, or anywhere on earth.

The GISS Temperature system is a complex system that strives to determine the average temperature of the earth in a unique way. This system was developed in the late 1970s and early 1980s to support single dimension climatic models. Therein lies the systems "Achilles' Heel," though there are a number of additional serious problems inherent, and interjected into it.

To help you understand this topic, we will first review for you how the temperature has been determined since the advent of the first thermometer. Then we will introduce what the "GISSTEMP" system is, how the need for this system is justified, and how it works.

My first consciousness of temperature, dates back to approximately 1947 (long after thermometers were invented), just before I turned four years old. It involved a large round aluminum disk with a glass face plate, behind which was a Coca-Cola advertisement. A large red arrow and numbers were spread over the top 2/3 of the white face behind the glass. This apparatus was placed on the outside of the North wall of the small building that housed the families "outdoor plumbing" (no I do not long for the good old days)!

For 10 years whenever we were hauling hay or working the livestock in the corral we could see that the mid-day temperatures were in the mid to upper 90's. Or when we were picking and hauling apples to the storage building in mid-October in a cold rain, we could tell if the temperature was below freezing. In both extremes, I thought that was not a very accurate thermometer, even though, when I compared the temperature shown there, to what was reported by the radio station 22 miles to the North, they were usually within two degrees.

That system was vastly improved in 1957 when we purchased a recording high/low thermometer and placed it in a wooden apple box nailed to the North side of the power pole that held the transformers for our home's electrical supply.

Over the next 60+ years the temperature systems involved in my life and life's work evolved to its present state. I can quickly use my cell phone to address the appropriate source and find out what the temperature is in Tagbilaran, or Mabini on Bohol Island in the Philippines, or Asahikawa on the Island of Hokkaido Japan or most any other location on earth.

If it is 3:15 a.m. on a cold spring morning and I want to know what the air temperature is in the orchards in southern Utah County (Utah, USA), I can use my computer or cell phone; Go to the Utah State University's Climate Center web site and address any one of the 20+ weather stations spread throughout the 10,000 acres of commercial orchards, in the southern portion of the Utah Valley from my home. A home, which is 60 miles and one 12,000' mountain away. The same thing applies if I want to check conditions in the Caldwell Idaho, or Wenatchee WA areas (different web sites, but easily accessed on my "favorites" list).

I can immediately see what the current air temperature, dew point, relative humidity, leaf temperature, bud temperature, wind speed, wind direction and the status of several other environmental factors are. In addition I can see a plot of the last three hours of air temperature, windspeed, and dew point and can access the historical data for that site (on these systems the readings are taken every 10 or 15 minutes, and are accurate to within 1/10th of a degree).

There are a myriad of other choices that are available to do such diverse things as monitor the skin temperature on the exposed vs. shaded side of an apple and other functions. All that is required is to define the need, determine the best choice out of the many available, buy it, install it, and monitor it (if what you need is not available you can make it).

For a few extra dollars I can monitor it from the easy chair in my living room, or from the top of what ever mountain I may be on. There are still problems in many areas where I like to be. If I am in the lower portions of some deep, steep canyon, my phone reception is not adequate, but only because I am too frugal to invest in the satellite phone necessary.

What I learn from these systems is expressed as a numerical value which has a fixed meaning, for example, the temperature is 27.2 degrees, the dew point is 25 degrees.

With that brief history and assessment of the state of the temperature monitoring systems we use, let us move on to the GISSTEMP system and anomalies.

What is GISSTEMP? How is the surface temperature of the earth determined and why is that relevant to the global warming/climate change doctrine? Bear in mind, this 'system' traces back to Dr. James Hansen around the time that the idea of global warming surfaced. Mr. Hansen is one of the Mindguards* of the global warming movement. (*See Appendix 1: p. 260 for the significance of that term). Let's take a look at the GISS Surface Air Temperature Analysis System.

But first, we must make sure we understand what an "anomaly" is. The GISSTEMP system relies on the Global Mean Temperature for the period of 1951 to 1980 to serve as the "normal" or reference period. You can locate that period on the graph below, which is an example of how a plot of this system from 1880 to 2014 looks.

Authors Note: The statement about the purpose of the 1951-1980 Global Mean Average Temperature was originally key to the system. In recent years there has been a tendency to use other "normal periods" and even edit the data for this reference period. You will see some examples in this work. Note: A change in the reference period (what is "normal") changes the results in this system. Also, a change in the temperature monitoring sites, their numbers, locations, and location characteristics can also change the results of the system. Such changes have been made, particularly when new datasets were adopted in 2011 and 2019. In this work, you will see how such changes have consistently increased the systems estimates of the warming of earth's surface (a highly questionable result).

From: https://earthobservatory.nasa.gov/world-of-change/DecadalTemp

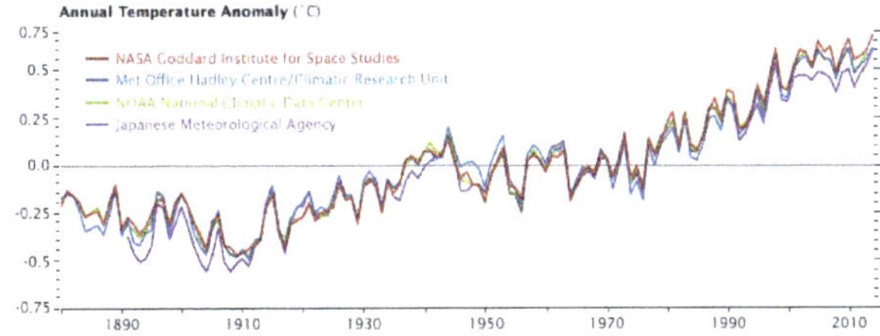

(Some of the NASA graphs that appear in this work were downloaded in 2018 & 2019 - they are

difficult to find currently.) Concentrate a moment on the period between 1950 and 1980. This was a period when the average temperature was relatively stable. But notice that it did fluctuate up and down. In 1956 for example, the "Anomaly" was -0.19^{0}C. The next year it was 0.05^{0} a change of 0.24^{0}C. So the "Anomaly" refers to the "difference" between the reference period global mean average temperature and the actual annual global mean temperature. In this instance the surface temperature went from being abnormally low, by -0.19^{0}C, to being abnormally high by 0.05^{0}C in a single year, then dropped significantly 8 years later.

That represents slightly more than 1/4 of the claimed 2016 record anomaly. According to NASA data at <Nasa CO_2 concentration 1850-2011.webarchive>; the 1956 atmospheric CO_2 concentration was equal to 314.3 ppm and the 1958 concentration was 315.34 ppm. This means that while the global temperature anomaly increased by 0.24^{0}C (equals 24.5% of the record 2016 claimed anomaly), the CO_2 concentration increased by only 1.04 parts per million. Notice and remember the frequency and degree, of the fluctuations of the anomaly.

Rhetorical Questions: Is the CO_2 concentration of the atmosphere closely correlated to the changes in the earth's surface temperature? What causes the highly significant rapid increases and decreases in the anomaly?

It is impossible to overstate the importance of this system, to what is, the global warming/ climate change doctrine/dogma. We will examine it in some depth, so; grab a little (or a lot of) caffeine and read on. To see the entire document being discussed go to: <https://data.giss.nasa.gov/gistemp/faq/abs_temp.html> Updated 2019-02-12:21:26.

We will discuss here only 3 of the Questions and Answers that appear in this document. Read this carefully and you be the Judge: Does the reasoning behind this system make sense or justify the system?

Question & Answer 1: Q. "What exactly do we mean by SAT?" (Surface Air Temperature)

A. "I doubt that there is a general agreement how to answer this question. Even at the same location, the temperature near the ground may be very different from the temperature 5 Ft. above the ground and different again from 10 Ft. or 50 Ft. above the ground. Particularly in the presence of vegetation (say in a rain forest), the temperature above the vegetation may be very different from the temperature below the top of the vegetation. A reasonable suggestion might be to use the average temperaturc of the first 50 Ft. of air either above ground or above the top of the vegetation. To measure SAT we have to agree on what it is and, as far as I know, no such standard has been suggested or generally adopted. Even if the 50 Ft. standard were adopted, I cannot imagine that a weather station would build a 50 Ft. stack of thermometers to be able to find the true SAT at its location."

The above answer may be a little surprising to you. You may have thought that you knew or could know what the surface air temperature (SAT) is. You probably also thought that your friends, family and others generally had some basis to understand what the temperature is or will be, after all the weather is one of the leading topics of discussion (and 1 click away on your device). With that in mind; Are you surprised by the above answer?

Did you think there was a workable system which had been generally adopted? Am I the only one who checks my device to see what the temperature and weather conditions are outside before we send the grand-kids to school?

Can you find out what the surface air temperature is by selecting your weather app and having it take you to the current temperature at or near your location? Does the fact, that the temperature you will see on your device will not be accurate to 0.1 degrees make it useless? Or, does the fact that; under a tree or on top of a very tall ladder, the temperature may differ from what you are told on your device, make the information useless? By the way, my granddaughter understands that under a tree the temperature will almost always be cooler than out in the sun, and she has known this since the time she could walk. That does not mean that my device, the information it can access, and my weather app is useless.

Does the following quote from the above answer ring true to you? "To measure SAT we have to agree on what it is and, as far as I know, no such standard has been suggested or generally adopted." We live our lives with what we know and learn on a daily basis about the temperature, and weather. It is a significant contributor to how we conduct ourselves. Am I the only one who thinks this quote is utter nonsense? We have a system, and it has been generally adopted (world wide) and is understood and very useful

Authors Note: The US National Weather Service has had very specific standards as to how air temperatures are measured for over 100 years. The locations of thermometers, the height from ground level, the type and color of enclosure they are placed in, how they are read, when they are read, and how the data is recorded has also been specified and largely followed meticulously.

Other national and international weather networks have adopted similar standards.

If I am going hiking in the mountains I check the temperature (current and forecast) to determine what gear to take, how much water to carry, and even the type of sunscreen to take along. If I am going to the Skyline of the Wasatch Mountains or to climb Kings Peak or Mt Nebo, I check the temperature forecast, and even if I can't find the exact temperature for the exact place, what I can find is extremely useful.

By the way, the suggested method of what it would take to determine the true SAT, you know, the stacks of thermometers, etc., would still only be "accurate" until a more complete set of stacks of thermometers of greater accuracy, height and more complete distribution was installed. (Is the pooch chasing his tail on this one? Hold your answer to that question until we examine the error factors involved with the GISSTEMP system.)

If you, like I, found the first answer a little incredulous, let's move on to Question and Answer 2: (Note the question numbers are not the order of questions in the document, merely the ones I choose to address in this discussion.)

Q. If the reported SATs are not the true SATs, why are they still useful?

> A. "The reported temperature is truly meaningful only to a person who happens to visit the weather station at the precise moment when the reported temperature is measured, in other words, to nobody. However, in addition to the SAT the reports usually also mention whether the current temperature is unusually high or unusually low, how much it differs from the normal temperature, and that

> information (the anomaly) is meaningful for the whole region. Also, if we hear a temperature (say 70°F), we instinctively translate it into hot or cold. But our translation key depends on the season and region; the same temperature may be 'hot' in winter and 'cold' in July, since by 'hot' we always mean 'hotter than normal',' i.e., we all translate absolute temperatures automatically into anomalies whether we are aware of it or not."

Authors Note: You may need to read this answer a second time. I did because I find some of it to be ridiculous, for example: "The reported temperature is truly meaningful only to a person who happens to visit the weather station at the precise moment when the reported temperature is measured, in other words, to nobody".

If you find the above quote does not fit you, if you find temperature readings that you were not visiting the weather station at the precise moment when the reported temperature was measured useful; welcome to the club of nobodies.

I would submit to you, that not even the authors of this system could believe the first sentence in the above answer. I base that statement on the fact that the entire GISSTEMP system depends on temperature readings from systems identical to, and in many instances the same ones that they assert, in this statement, yield data that is "truly meaningful only to a person who happens to visit the weather station at the precise moment when the reported temperature is measured, in other words, to nobody."

Were the authors of this system visiting every weather station from which data was used to determine the reference global mean temperature of the earth's surface at the precise moment when that reported temperature was measured? That would apply to every weather station used world-wide, every day for 30 years (1951 to 1980). This question would apply to all the weather stations involved in this system from 1880 to present.

They were not there, so their data is useless; their reference global mean temperature is useless. In short, if this statement is true the system is useless. Rhetorical Question: How has this statement survived the several revisions that this document has gone through during the last two or three decades? It requires a special kind of bias to reach such a conclusion.

Another portion of the above answer that is, on its face preposterous, is: "in addition to the SAT the reports usually also mention whether the current temperature is unusually high or unusually low, how much it differs from the normal temperature, and that information (the anomaly) is meaningful for the whole region". In this system "region" refers to the area described by a 1200 km radius circle from the point the temperature is observed and recorded (an area of over 1.745 million square miles).

If, on a bright sunny February day in Salt Lake City, Utah the anomaly is +5^0 can I drive to the same altitude in the Wind River Mountains of Central Wyoming and count on the anomaly being +5^0? It is in the same region. The same question could be asked about the Alta, or Park City Ski areas just a few miles east of Salt Lake City.

Hint; If you believed this part of the above answer may I suggest that if your destination is any of the 3 sites named as possible destinations, take along the cold weather gear. If the wind is blowing at the higher altitude you may experience an anomaly of -20^0 with an additional huge wind chill in spite of the fact that all three are in the same “region”. If, on the other hand, there is an inversion and the anomaly in Salt Lake City is -5^0, you may be skiing in your shirtsleeves at Alta or Snowbird when the temperatures are well below freezing in Salt Lake City.

For over 50 years I have been involved with and dependent on plant, insect and disease development models. These models were driven for the first 20 years by temperature data that was obtained from National Weather Service (NWS) Hi-Lo thermometers.

Over the last 30 years the data has been obtained by automated temperature collection systems operated by NWS and various university climate centers (and from systems owned and operated by agricultural enterprises). I have not been present at the precise moment that the individual readings were recorded (most of the current systems record the data every 10 or 15 minutes). I have found this information very useful, as have the agricultural growers and managers, that have paid me well, to teach them to use and interpret the models, and the systems that collect the data for them.

In the final round of the ridiculous statement contest, I think that the following quote should do well. “But our translation key depends on the season and region; the same temperature may be 'hot' in winter and 'cold' in July, since by 'hot' we always mean 'hotter than normal', i.e., we all translate absolute temperatures automatically into anomalies whether we are aware of it or not”.

The above quote is partially true but the latter portion is patently untrue. You be the judge;

“since by 'hot' we always mean 'hotter than normal', i.e., we all translate absolute temperatures automatically into anomalies whether we are aware of it or not”. When you say or think that the temperature is hot, is it always in the context of hotter than normal? If it isn’t, the first part of that statement is, by your admission, untrue, and the fact that it is untrue strongly indicates that you do not automatically translate temperatures into anomalies.

As an example of what I mean by the last paragraph. My family and I spent a wonderful summer on a friend’s property along the Virgin River east of Springdale Utah (next to Zion National Park, USA). Beginning on June 6, through mid September we experienced 92 days when the high temperature ranged from 102^0F to 114^0F.

With beautiful red sandstone ledges 500’ to 1000’ high surrounding us to absorb and reflect the heat, it was always hot outside (except from 11 pm to 5 am). It seems that I and the other members of the family expressed that it was hot many times a day, but we realized that, in this location, at this time of year it would be hot. Equating “hot” to “always means hotter than normal” is a stretch in several dimensions.

I could go on and on, but there is one additional question and answer (the last one) in this document that reveals a very important fact. So here is that, "Question and Answer".

Q. What do I do if I need absolute SATs, not anomalies?

> A. “In 99.9% of the cases you'll find that anomalies are exactly what you need, not absolute temperatures. In the remaining cases, you have to pick one of the available climatologies and add the

> anomalies (with respect to the proper base period) to it. For the global mean, the most trusted models produce a value of roughly 14^{0}C, i.e., 57.2^{0}F, but it may easily be anywhere between 56 and 58^{0}F. and regionally, let alone locally, the situation is even worse."

The above answer has two very revealing parts. <u>Rhetorical Question:</u> Is the following true? "In 99.9% of the cases you'll find that anomalies are exactly what you need, not absolute temperatures".

If you need to determine whether or not you or your children need to take a sweater, or a coat, when you leave the house, would you rather have, a temperature (for example the current temperature is 42 degrees) or the anomaly (for example "it is hotter than normal, normal, or colder than normal" since it would always be hotter than normal, normal, or colder than normal)?

In terms of numerical expressions of an anomaly, how useful would it be to know that the anomaly is 0.98 degrees above the 1951 to 1980 global mean temperature. Would that help me decide if I needed to insist that the grand kids wear a coat instead of a light sweater?

For example during the month of October this year (2019) there were 2 episodes of extremely cold temperatures for several days. Temperatures ranged from 20^0 to 35^0 below normal at my home site. What use would it be for me to know that the temperature was going to be "colder than normal," when the normal low temperature for those dates would be in the lower 40's. I would have felt good about my grandkids wearing a light jacket if I thought the temperature was "lower than normal" but the picture was much different when I knew the outside temperature was 4^0F at my home and -4^0F in Fairview where they go to school.

An example of when an absolute numerical temperature is necessary and an anomaly from the GISS system or other anomaly source would be totally useless, often occurs when apple trees are in bloom. Several of the most popular and highest priced varieties are prone to a bacteria caused disorder called fireblight (a disease that can kill a significant portion of the trees in a planting if a single necessary spray application is missed).

This disease is very sensitive to temperature. Below a certain temperature, 59 to 60 degrees, it is not infective. Above this rather narrow temperature range the disease is infective and can be spread very rapidly by honey bees as they pollinate the flowers.

Using a model of the disease behavior at various temperatures and the short-term temperature forecast we know when a control spray is required. This model, like every other model we use in disease and insect control is dependent on accurate temperature numbers. Control measures are much more effective as a preventative, with their effectiveness dropping very quickly once the threshold temperature is reached.

In many orchard districts in Washington, Oregon, Idaho, Utah and Colorado the average mean temperature during the bloom period will be below the threshold temperature. The actual temperature may be well below the threshold (2^0 - 5^0) then the next day be well above the threshold.

In the above instance if we were to rely on an anomaly instead of the actual temperature several bad things would happen. They are:

1. Every time the temperature forecast was for temperatures "above normal" growers would be required to spray, and keep spraying every 4th day as long as the temperature forecast remained "above normal". This could require 5 to 9 spray applications during the bloom period every year which would be very costly. *Note: To use an anomaly, e.g., -3°F to determine whether an application of a control material was necessary, the grower would be required to go through several additional steps or mental calculations. First they would have to find the normal temperature; Second they would have to subtract 3°F; Third they would have to compare the resultant temperature to the threshold or critical temperature; Fourth they would then be able to make the decision as to whether or not to apply the control application. Using the absolute forecast temperature for the next 12 hours they would compare that to the threshold or critical temperature, then make the decision; Much more efficient and time saving when a critical decision is required.*

2. If the grower did not go through the extra steps, in addition to the extreme cost factor, they would find that they would rapidly select resistant strains of the disease (strains of bacteria resistant to the spray material being used). The purpose of using the model is not just to lower expenses, it is a key part of resistance management, something which is essential in preserving the effectiveness of the limited number of effective control materials. Without good resistance management a significant number of foods would quickly become more expensive and eventually unavailable in the general market. The Irish famines which caused extensive population dislocations were caused by potato blight,

which can still cause severe damage if not controlled. Models and timely but limited applications of control measures are required.

3. Resistance to antibiotics used in agriculture or human health care can be selected rapidly, as we all know. My father passed away from an antibiotic resistant infection picked up in an operating room during what would have been, a successful open heart surgery. Within my peer age group, a common medical need is hip, knee, or shoulder surgery. An increasing percentage of these surgeries result in antibiotic resistant strains of infections showing up post operation. Currently there is only one antibiotic treatment for many of these infections which is a high, daily dose, over an extended period, (weeks or even months) of Vancomycin. For certain strains, when this material is gone, the only cure will be aggressive amputation (something that is unpredictable and undesirable).

There are literally hundreds of examples that could be given of this principle. The continued control of these diseases and insects is highly dependent on accurate, temperature data (Critical numerical temperatures, not an anomaly). Knowing that the temperature was going to be, or had been, cooler than normal, normal or warmer than normal would be totally useless. The assurance that, "In 99.9% of the cases the anomaly is exactly what you need" is rubbish, aged in a moist, warm environment, (conditions that usually result in the release of noxious gases) .

Finally, in the last sentence of the above quote which reads, "For the global mean, the most trusted models produce a value of roughly 14^{0}C, i.e. 57.2^{0}F, but it may easily be anywhere between 56 and 58^{0}F and regionally, let alone locally, the situation is even worse"; something extremely important is revealed.

The admission here is that the GISS anomaly system has an error of at least plus or minus 1 degree F, on a global basis and regionally, the situation (error) is even worse. What does this error mean? Briefly it means that any actual anomaly of less than 1.0 degree F (0.6 degrees C) cannot be judged to be different from the mean. It is said to be within the margin of error.

Further, if the anomaly exceeds 1.0°F, for example if it is 1.1°F only the 0.1° is outside the margin of error. Later in this chapter we will discuss this in significant detail. The statement that the SAT "temperature estimate," on a regional and local basis, is even worse means the error is greater than 1°F. There are additional errors from ignored bias factors and from errors interjected by such practices as changing raw data, homogenization of data sets, or LOWESS smoothing. With an anomaly system, an error of 1°F or larger is a real problem, unfortunately deemed to require additional data manipulation.

It is important that you note the date of the above NASA document. It was updated and released on 2019-02-12:21:26. That is February 12, 2019, 26 minutes into the 21st hour. Three months later NASA adopted a new data set, GHCNv4, indicating it is version 4 of a doctored series of data sets, which totally disregards this admitted fact (there is a +/- 1°F error). I know that sounds harsh and judgmental. You may wish to reserve your judgment until you have read how and why it is true. (This will be discussed later and I will remind you of this date, after all that is part of my job.)

At the beginning of this chapter I pointed out that this system is designed to support single dimensional climatic models. I referred to this fact, as the "Achilles' Heel" of the system. At this point I am sure you understand what that means, but just in case there is confusion let me state why this is the "Achilles' Heel".

In the first chapters of this work I discussed 4 environmental factors or variables which have greater impacts on the earth's surface temperature than the 1/25 of 1% of the atmosphere that is CO_2. Those variables are:

1. The suns temperature is known to vary and cause the intensity of the sunlight which reaches the earth's surface to vary. Since the 1940s solar activity has been considerably higher than at any time in the history of this force being monitored. While the variation is not large, there are significant interactions with other environmental variables which result in a multiplication of its impact, including energy stored in the earth's surface and waters. (See Chapter 1.)
2. The climate and temperature control of the earth's surface that results from the water that covers 75% of the earth's surface (See Chapter 3).
3. Heat from the earth's core in the form of volcanic activity, magma flows, thermal vents, convective heat movement through the earth's crust and the polar ice cap melt caused from these heat sources. This heat source, like the sun, increases the evaporation of water from the earth's seas which multiplies its impact in significant ways (See Chapter 4).
4. The greenhouse impact of particulate and aerosols in the atmosphere which has been increasing steadily and significantly over the last 100+ years as populations have increased, and particularly since 1920 as

forest fires and land use conversion have drastically reduced the area of the earth covered by forests (See Chapter 5).

The warming influence of the above 4 forces have all been added into the temperature increase which is attributed too CO_2 forcing of the earth's surface temperature because of the "Achilles Heel" of a single dimensional modeling system. (See: Evolution of the Anomaly at:<fatalflaws.net>)

Each of the possible interactions between the above 4 forces and the minor greenhouse gases need to be considered. Intentionally lumping them all together under one moniker can only result in multiple errors (Fatal Fails).

The underlying assumption is: Since there is a weak correlation between the increasing temperature of the earth's surface and a rising atmospheric CO_2 concentration, all the warming is due to the rising CO_2 content of the atmosphere. That is the forced conclusion of a single dimensional analysis. Nature is not a system of single dimension interactions!

I pointed out earlier how a small increase in solar intensity warms the surface temperature of the earths waters, causing greater evaporation. Increased evaporation results in greater movement of heat from the earth's surface into the mid-levels of the atmosphere. This energy is in the form of chemical "potential energy" in water vapor which is not subject to interception by greenhouse gases.

This energy by-passes much of the potential greenhouse impact of the minor greenhouse gases. There is also a synergistic interaction because of the ice cover loss over the Arctic Ocean caused by natural variation. Open water reflects less of the solar radiation than ice, with more of the heat being taken up by the water which further enhances the ice cover loss and delays freeze up the following fall.

In the third paragraph above I referred to the correlation between the rising temperature of the earth's surface and the increasing concentration of CO_2 as a weak correlation. Since 1880 there have been periods when the correlation was negative, other periods when it has been extremely weak and periods when it was moderate. (Remember always: Correlation does not prove causation.)

I preformed a correlation analysis between the NASA data on the CO_2 concentration (NASA CO_2 concentration 1850 to 2011.webarchive) and the Anomaly (NASA Anomaly Data 1880 to 2018.webarchive). Below is a graph showing the Anomaly for the years of 1880 to 2014 (yes you saw it a few pages earlier). Take a close look at the start-point of the lines in 1880. The start-point is equal to -0.16^0C, meaning that the global mean temperature that year was -0.16^0 lower than the 1951 to 1980 average global mean temperature. The 1951-1980 period (30 years) is the reference period or "normal" for this system.

In 1880 the CO_2 concentration was 291 ppm. By 1980, one hundred years later it had reached 339 ppm, a 48 ppm increase. That represents 46% of the total increase between 1880 and 2014.

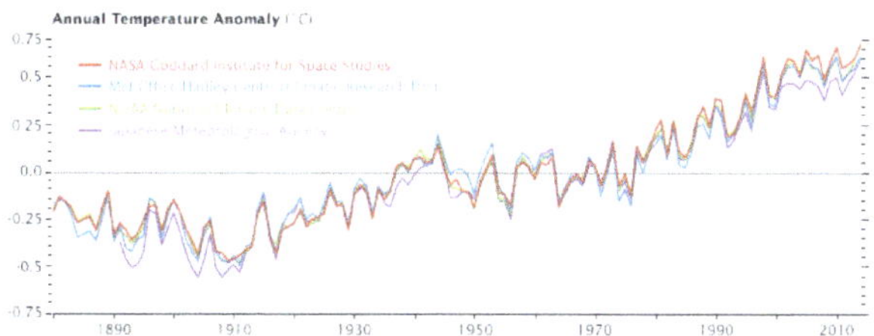

As you can see from the graph above the 1980 global mean temperature was approximately

0.18^{0}C higher than the 1880 figure, even though the CO_2 concentration had increased by 46% of the total increase in the time period represented by this graph.

You will note that between 1880 and 1912 the temperature dropped by -0.3^{0}C while the CO_2 concentration rose by 11 parts per million. These are not characteristics of a strong correlation. A correlation analysis of the change in the anomaly vs the change in CO_2 concentration is actually negative for the period 1880 to 1930 (50 yrs).

During the 1951 to 1980 "normal" period the Correlation Coefficient (r value) was equal to .52 and the Coefficient of Determination (r^2 value) was only .27. This means that during the normal or reference period a maximum of 27% of the change in the earth's surface temperature could be attributed to changes in the CO_2 concentration of the atmosphere.

When you combine the negative correlation of the first 50 years with the weak correlation coefficient of the "normal period" it yields a very weak correlation. During that 100 year period the fluctuations in the atmospheric temperature were almost entirely due to factors other than the increasing CO_2. (if that went over your head I will explain it fully in an understandable manner in Chapter 11, the conclusions chapter).

Indeed, the above graph illustrates how forces other than CO_2 have been the major factors in control of the earth's surface temperature. The reason I did not include the post 1980 period in the above analysis was because of the major changes that have been made in the data that covers that period. We will discuss that in some detail in the next chapter. Should we move on?

I would like to, but there is one more thing that needs to be illustrated. It shows in graphic terms how internally inconsistent this system can be. Take another look at the graph above (p.177), noting the section centered on 1960 that forms a "M" like pattern. Further, note that the rise discussed above is followed by a small dip and then a major decrease in the mid 1960s. This shows a 0.24^0 rise, a small dip and then a 0.26^0 fall in the global annual anomaly. Rhetorical Question: This time period corresponds to solar cycle 19, the most active cycle in recorded history. Was it a mere coincides that this highly significant increase, plateau, and then drop of the anomaly occurred at the same time?

Authors note: The chart of NASA data from which the above was obtained has a second column entitled "Lowess (5)" indicating that the Lowess smoothing statistical technique, using a 5 year rolling average was used on the data in the 2nd column. This technique is used to "smooth" the data so the variations are not as pronounced (the claim). It actually hides the fact that there are significant variations that are caused by other forces. In this case the range (of the Lowess smoothed data) for the 10 year period of 1956 to 1965 is from, -0.05^0 to 0.03^0 for a net range of 0.08^0 which is less than 1/3 of what the actual data show (actual data = 0.24^0)C.

You will see in the next chapter how the use of Lowess smoothing, long term rolling averages and other techniques are being used more frequently.

The bottom line is that your trust of data and information from even the most trusted sources needs to be cautious and qualified. Some of these methods are represented as being for convince, when they are actually deceptive.

Authors note: In this work I have used the historical temperature anomaly graph on page 180 several times. I refer to this graph as the "Berkeley Earth Graph". The source of this graph is an article entitled "Thorough, not thoroughly fabricated: The truth about global temperature data" in an ars Technica publication @ (https://arstechni-

ca.com/science/2016/01/thorough-not-thoroughly-fabricated- the-truth-about-global-temperature-data/). This graph closely reflects the methods used in the formulation of the 2011 and 2019 revisions of the GHCN data set series (GHCNv3 & GHCNv4) used to determine the Global Mean Temperature Anomaly.

When you see this historical plot of the global mean anomaly in this work please note how the 3 different data sets shown on the graph do not apparently need very much modification until approximately 1970. Following 1970 you will notice a very significant, gradual divergence of the raw data line (in red), the adjusted data line (in blue) and the fully corrected and homogenized line (in green). In other words, to formulate the fully corrected and homogenized data (currently termed "the unadjusted data"), which is used to determine the anomaly has required more and more "corrections" of larger and larger magnitude as time passed between 1970 and the current time. Ask yourself why is this, every time you see this graph.

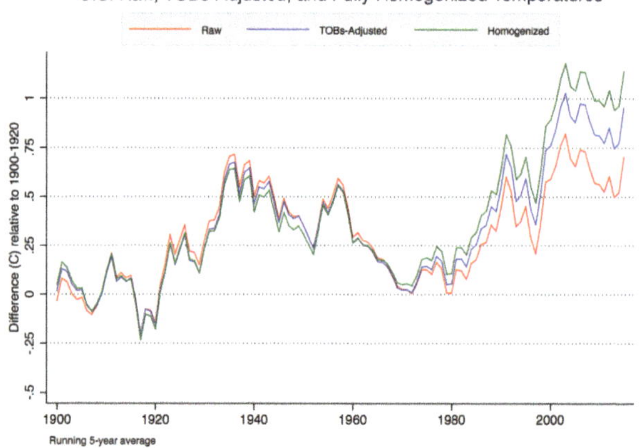

Chapter 10: How Much Has the Earth's Surface Warmed?

<u>Synopsis:</u> *Next, we will explore the fact that the raw data numbers are being changed, and in spite of the fact that there are millions of data points involved, the overwhelming*

preponderance of these changes increase the "warming trend". The amount of data involved, the types of possible errors, the time course of events and other factors would dictate, that statistically, the cooling errors should come close to canceling out the warming errors. Such is not the case, in fact nearly 50% of the claimed increase in the earth's surface temperature is attributable too outlandish data manipulation.

We will discuss the reasons, and illustrate, with their own documents the absurdity of the practice of changing the raw data, and how, on the face of their documents, their justifications are proven false.*

**Authors Note: In this paragraph and in several other places I write that the raw data is being changed. In a technical sense that is not correct. The raw data is not being changed, <u>it is just not being used.</u> What is being used, is the "unadjusted data," which is the raw data after it has gone through a 5 step series of "corrections". There have been several versions of the "unadjusted" data, the GHCNv1, GHCNv2, GHCNv3, GHCNv3.1, GHCNv3.2 and now the GHCNv4. I will write more about this later.*

A well designed and executed scientific experiment or system would not have the problems this one has. For example the ocean average mean temperature is determined by several entirely different systems. Each have their problems but produce data. The data from these different systems is "homogenized" into one data set. This occurs after corrections, adjustments, guesses (extrapolations), and estimates of errors brought about by the mixture of equipment, the variations in the depth of sampling, spacial distribution of sampling sites and calibration errors. This results in a numerical environment that is ripe with bias.

In 2003 a world wide buoy system involving approximately 3,000 sites was placed on line. In 2005 a peer reviewed report was published involving 20+ authors and coauthors. The thrust of this initial report was that in the 2 year time period (2003 - 2005) the temperature of the earth's oceans had dropped by 0.11°C, which represented 20% of the historical, estimated, increase in ocean temperatures. That result shook the global warming/climate change community.

Six months later a number of the original authors and coauthors published a report which maintained that the problem had been, a large group of these sites' sensors had a calibration problem. They made a mathematical recalibration, fixed the problem (changed the data) and the system has operated since that time, but instead of being a trusted, accurate system it is only part of the homogenized stew.

This extremely expensive system apparently never was actually physically tested and calibrated. My mentors would term that, "A freshman mistake". They would say, if you don't take the time and care to do it right the first time you will have to take the time and care to do it over. In this case, they were wrong. A quick fix involving a thick, cooked mixture of milk, cocoa powder, confectioner's sugar, and water (fudge) was used to "fix" the problem.

The motivation is clear. If you don't fudge the numbers the weaknesses of the dogma are exposed. Rhetorical Question: Is changing the data, using long term rolling averages and homogenizing data sets an indirect admission that the global warming/climate change doctrine is fatally flawed leading to fatal fails? (An "advantage" to this situation is, if the errors are ill defined and unknown, you can use the numbers in ways that prove whatever you want, or you can change the numbers to reach the desired ends.)

Photo: Enjoy the beauty of a cold sundown at 9,780' elevation

Chapter #10: I apologize for the following 'trip into the technical weeds'. It is necessary and I will work to make it as painless as possible. If you will work a little you will begin to realize what a house of cards this system is.

Examine the chart below of the Annual Temperature Anomaly (You have seen it twice before but we are going to look at it in a different context). Note that the last year represented on this graph is 2014 and that all 4 lines indicate the annual anomaly was between 0.7^0 to 0.75^0 C for that year. (Note that the range and values had been nearly flat from 1998 to 2011.)

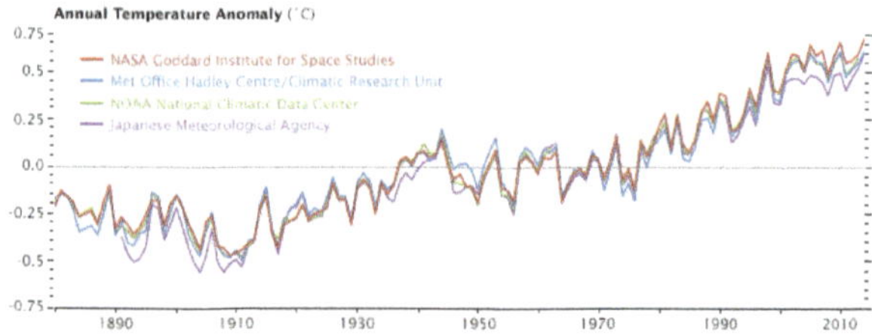

In Chapter 9, I discussed the last question and answer from the document that describes the GISSTEMP SAT system. The last sentence in this document reads: "For the global mean, the most trusted models produce a value of roughly 14^{0}C, i.e., 57.2^{0}F but it may easily be anywhere between 56 and 58^{0}F and regionally, let alone locally, the situation is even worse." I would like to illustrate the importance of that admission as follows.

The graph below represents the Global Mean Estimates based on Land Data Only. Later in this work I will discuss why the land data are much more reliable than the land data plus the ocean temperature data.

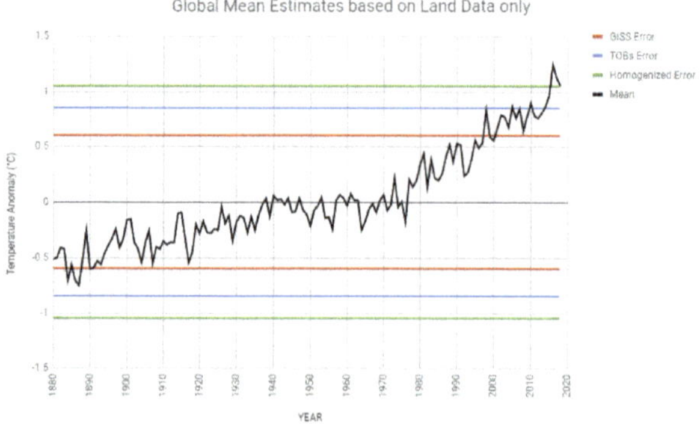

Note the following on the graph above:

1. The 1951 to 1980 period which represents the normal surface air temperature (the 0 line).

2. The positions of the two red lines at -0.6 and +0.6. What the statement above which reads Quote: "but it may easily be anywhere between 56 and 58°F," admits is that, either position of these red lines, or anywhere between, could, in fact, be the mean global surface temperature for that 30 year period.

3. Note the positions of the two blue lines representing a *0.25^{0} C "correction". These lines indicate what the confidence interval would be if the major data changes that have occurred were considered errors.

4. Note the positions of the two green lines. The *0.2^{0} C "correction" that resulted when a number of data sets were combined (homogenized) is illustrated by the space between the blue and green lines. (*From Berkley Earth Graph p. 180; 2010 values vs the raw data value of 0.5^{0}C)

Surprising isn't it? This graph illustrates what I mean when I say that the error factors involved make any conclusion based on this system of tenuous value. There is no point (prior to 2011) where one can say, with certainty, that the anomaly is greater than the admitted error. (The steep rise following 2011 is the result of the "warming effect" that occurred when the CHCNv3 data set was adopted.) Now examine the graph below and the explanation that follows.

The graph (p. 186) shows what happens if the 1951 to 1980, global mean surface average temperature, actually lies at the lower limit of the admitted error. Now you see why I wrote that this system (scheme) can't be relied on. It is proven to be suspect on its face.

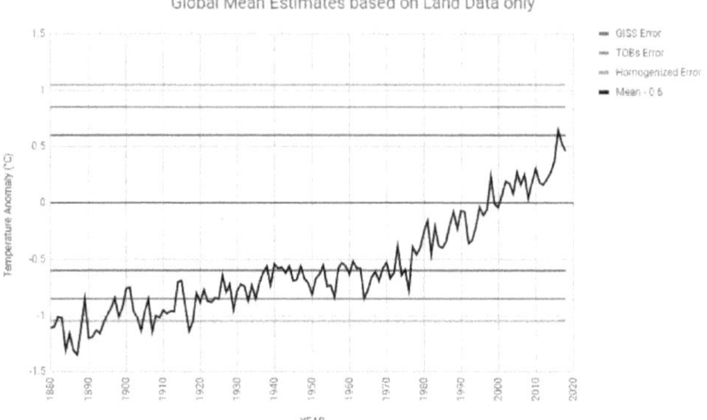

You now have a good view of another aspect of the "settled science" behind the global warming/ climate change doctrine (this is one of a number of reasons I use the term doctrine to distinguish the theory from the facts). This calls into question the very foundation of the theory. You ask; How? Pay attention to the following.

The admitted error in the GISS system "global average surface temperature" is plus or minus 1 degree F (0.6 degrees C). Added onto that should be an additional error factor, derived from the regional and local errors as it is extremely unlikely that they are perfectly balanced across the world. They must be significant considering the language used. Additional error factor(s) exist as we will discuss later (these errors are not included in this phase of the discussion).

In addition to the admitted GISS Error you will note the two blue lines which represent the TOBs Error. The difference between the red and blue lines represent the magnitude (0.25^0 C) of the changes in the raw data when it was converted to the GHCN "unadjusted" data. The space between the blue and green lines represents the warming effect which occurred when the data homogenization scheme was adopted.

In 2011 the use of the GHCNv3 data set began. This data set included all the previous adjustments made in the "unadjusted" data set including data set homogenizations. There also was inclusion of data from a significant number of additional weather stations. The impact of the adoption of this data set was a vastly increased slope of the curve between 2011 and 2016. Do you believe in coincidences?

You will note that the error factor introduced by the combination of changing the raw data and the homogenization of data sets adds an additional 0.45^{0}C error, making the total error plus or minus 1.1^{0}C. This extends the confidence interval well beyond even the most outrageous estimates of the Global Surface Temperature Anomaly.

Hopefully you comprehend that this renders this system totally unreliable, and it is so on its face. <u>Rhetorical Questions:</u> Do you now have even greater confidence in the system we use as we live our lives? Is the system useful only to the vast club of nobodies described above? Isn't it nice to click on an icon and find what the current temperature is and what the forecast temperatures will be (with a very reasonable degree of certainty)?

I wrote above that the Global Mean Temperature based on Land Data Only, was more reliable than the global land and sea data. The following will form the groundwork for me to explain why that statement is true. Be patient, we will get there.

Additional Error Factors: The first paragraph of the above quoted and discussed NASA article contains the following sentence. "Indeed, we have shown [Hansen and Lebedeff, 1987] that temperature anomalies are strongly correlated out to distances of the order of 1000 km." In other words, in the GISS system each observation point's temperature anomaly is spread over a circle with a radius of 1000 km. (That radius has actually been expanded to 1200 km in the current version of GISS.)

The line quoted in the previous paragraph indicates "temperature anomalies are strongly correlated out to distances of the order of 1,000 km". This claim is absurd when you consider that:

1. This claim describes a circular area with a radius of 1,000 km, covering an area consisting of 3.14 million square kilometers (km^2). (They actually use a radius of 1,200 km currently in the system which equals 4,521,600 square kilometers or 1,745,798 sq. mi.) Based on the above publication the following describes how it works. We will look at 3 NWS weather stations; one in Spanish Fork Utah (USA); one at Salt Lake City (airport); one in Logan Utah. These weather stations are all at elevations between 4446' and 4498'

    a. At each of these stations the temperature data is recorded and from these records, the monthly mean temperature anomaly is determined and assigned as the monthly anomaly across the 1,745,798 square miles. (Yes, those determining the anomaly, are among the "nobodies" who find these temperature observations useful; Even though they were not present at the instant the temperatures were recorded.)

    b. That monthly temperature anomaly (and the 11 others for that year) is then used to determine the annual mean temperature anomaly for the 1,745,798 square miles within the 1200 km radius. That circle for each of these stations reaches to Edmonton Canada on the North, Eastern Kansas on the East, into Mexico on the South, and to approximately 200 km west of SanFrancisco.

c. The annual mean temperature for those three sites, and the other 28,000 sites world wide is then used to determine the average surface temperature of the earth. That temperature is then subtracted from, or added to, the 1951 to 1980 mean global temperature to give the annual average global temperature anomaly.

d. If that is difficult to understand, consider that every weather station within the 1200 km radius of these 3 stations does the same thing so the temperature data from each of these stations results in a redundant layer, with different values, stacked on top of other redundant circles. So the monthly mean, the annual mean, and the annual mean anomaly for these 3 locations is actually an average of temperature observations at those stations, and every other station that lies within a 1200 km radius circle. If that sounds like a mess go to (Hansen and Lebedeff, 1987) and look at Fig. #1.

As an exercise to examine how this system is used (and abused) let's look at the concept of an urban heat island effect. Before global warming, it was recognized, measured and considered but now is not recognized or is hidden with a "slight of the calculator" trick. If it is properly considered it would mean a smaller "carbon dioxide forced" anomaly. So a number of "experts" took a look at whether or not there was such a thing. They did this using the GISS system and pulled the urban stations for a specific city out of the data set.

To fully understand what I am going to write next you should go to the original article (<Hansen and Lebedeff, 1987>) and look closely at Figure #1. Assuming that few will actually do that, let me describe what you will see. Remember that every weather station involved is represented by a black circular line with a 1200 km radius. In the 1870s, the first quadrant of Figure 1, there were a few scattered circles in a few places. By 1960 the overlapping circles turned the land masses and a few hundred km of the surrounding seas into a black mass. The Arctic, Antarctic, and a few areas in the centers of the oceans are exceptions. Now you are prepared for me to tell you how the conclusion about the urban heat island affect mentioned in the above paragraph is so predictably flawed.

Using Japan as an example lets see how this system could be abused to "prove" there was no such thing as an urban heat island effect, or to diminish its magnitude. We will choose the city of Tokyo which occupies the northern portion of the Tokyo Plane. This Plane is also the location of Kawasaki and Yokohama cities. Together the total inhabitants of this area, roughly 70 miles north to south and 30 miles east to west, is a little more than 38,000,000 people representing approximately 1/3 of the countries population.

A few other essential statistics would be:

1. According to the Japanese Climate Ministry there are 931 automated temperature sites monitoring the temperature using electronic thermometers to measure and record the daily maximum and minimum temperatures. I could not find the number

of these sites that are in Okinawa Prefecture which is separated by some distance from Kyushu (the southernmost of the 4 major islands). Given the size of that Prefecture (quite small) I will make a convenient, educated guess that there are 31 of the 931 total sites in Okinawa and 900 on the 4 main islands.

2. According to measurements made of Japan, (on Google Earth Pro), from the northernmost point of Hokkaido (northern island) to the southern limits of the Tokyo Plane is 1167 km. From the southern most point on the island of Kyushu (the southern island) to the northern limits of the Tokyo Plane is 1033 km. That means that every weather station on the 4 main islands of Japan overlap the entire Tokyo Plane. If you pull the data of every weather station in the Tokyo Urban area, you still have 872 overlapping circles that each represent a different weather station from all over the 4 main islands. The temperature data for each of the 872 unique temperature observation sites are spread over and averaged into Tokyo's temperature.

3. The next step in "proving" that there is no such thing as an urban heat island effect, is to compare the temperature anomaly before you pulled the Tokyo data and the anomaly after you removed that data from the data set. So now you have a temperature anomaly from which 3.1% of the total sites have been removed, but 96.9% of the original data is still in place.

4. The 3.1% of the sites you pulled from the analysis represents only 1/28 of the total stations which contribute temperature

data to the Tokyo Plane average temperature. What if the 3.1% of the total stations actually had a mean monthly and annual temperature that was two degrees F (1.2^{0}C) above the monthly and annual mean of the other 872 stations? A two degree elevated temperature, divided by 28 is equal to 0.07 degrees, which is within the experimental and rounding errors so the conclusion is: "There is no urban heat island effect". Congratulations you have just created an artifact! Now you can use it as an "Artful Lie".

5. Number 4 above was the initial "solution" but after a significant amount of questioning the position of: "There is no urban heat island effect" was left as the domain of Wikipedia and related sources. The datasets were modified by dropping the intercity sites and substituting a nearby urban site, a solution that solves nothing because the entire plane is densely populated.

At this point, using Japan as an example, I will defend the statement I made earlier that the Global Mean Temperature, based on Land Data Only, is a better indicator than the Global Annual Temperature Anomaly (this answer also relates to the discussion immediately above). The 931 weather stations in Japan, would each be the center of a 2400 km diameter circle where the temperature (anomaly) of that one station would be imputed as representative of the anomaly of the entire circle (over 1.47 million square miles).

At Honshu's widest point (on the Tokyo Plane) and Hokkaido's widest point (from Otaru on the West to Abashiri on the East) the width is just short of 400 km. That means that most of the area under each of the 931 circles that cover

Japan lies over the pacific ocean on the East, and the sea of Japan, extending across Korea or into eastern Siberia on the West. The air temperatures above the oceans tend to be significantly cooler than the air above land, particularly at the northern latitudes.

This means that this system creates a highly significant warming bias (the difference between the land air temps, and the sea air temps) over an area of ocean that is, at the minimum, 3 times the surface area of the nation of Japan. If you examine closely Fig. #1 of the above reference you will find that there is a large percentage of the ocean which falls under the land based circles. This temperature spreading results in three significant warming biases.

1. The numbers of these land based observation points dwarfs the number of ocean centered sites (a warming effect).
2. Much of the ocean surface that does not fall under multiple land based regions is grossly under-represented. Almost 1/8 was not represented at all (not included in any circle during the reference period 1951 to 1980) and the ocean buoy system is widely dispersed and was more-so during the reference period (a warming effect).

3. In an attempt of correct this problem, data from the cooling water intake ports of ships, is homogenized into the dataset. This does very little to "rectify" the bias, in fact it increases it. The temperature of the water taken in for cooling systems of ships is warmer than the average sea surface temperature because:

    a. Most shipping occurs in the temperate and tropical zones and the coldest areas of the earth's seas are avoided as much as possible because of storm, and ice threats.

    b. The depth of sampling varies greatly depending on the load status of the ship, and how far under the water line the intake is, in other words, it is generally warmer than the surface where evaporation is occurring (Remember the 4 step experiment we preformed to help understand what the "latent heat of vaporization" is?). These three facts result in a huge warming bias. This error, exaggerates the degree of warming, by artificially increasing the "measured" air temperature over much of the oceans.

Another significant error, which is hidden by the GISS method, is an altitude bias. If we examine the North and South American Continents we see that, beginning in Northern Canada and Alaska, south to and through the western U.S, through Mexico and central America, then the entire length of South America is a vast expanse of mountains.

The towns and cities tend to be located in the valleys thousands of feet below the tops of these mountains. Temperature monitoring stations are generally in or next to the cities and larger towns. (This trend of high elevations not being represented applies to other vast areas around the world.)

This means that the temperatures of the valleys are spread across the higher elevation areas, which are much colder, and the high elevation areas are not represented at all, or at least they are "under represented". This results in large areas being assigned temperature values significantly warmer than they actually experience. Let us examine what I mean with the example of Vail Colorado (the ski resort area).

Vail Colorado lies at an elevation of 8022', which makes it one of the highest elevation cities in North America. Assuming that its weather station is one of those involved in GISSTEMP, it lies at the center of a circular area with a radius of 1200 km. This area includes the heart of the Rocky Mountains from southern New Mexico, northward to the Wind River Mountains, Glacier National park, Southern Canadian Rocky Mountains and much of Montana. It also includes the Wasatch Mountains and most of the Great Basin areas of Utah and Idaho. To the East it stretches too, and a little beyond the Mississippi. In short it includes a tremendous area which is higher in terms of altitude, vast areas which lie to the north of Vail, and large areas which are significantly lower.

Vail is within 1200 km of a large number of weather stations, each of which contributes temperature data that is included in the redundant layers of data that are stacked upon (averaged into) its temperature. Many of these stations are located in lower, warmer areas which tend to bring the average temperature for Vail to a significantly warmer level than it actually experiences. That results in a significant warming bias being interjected into the data base, but that is only one of three, altitude related, warming biases.

There are vast areas in the 1200 km radius of the Vail "circle of influence" that are significantly higher and colder, on average. Many of these areas are 10,000' in elevation and a significant number of mountains range up to 13,000' and even 14,000'. These areas are assigned temperatures, and their surface temperature anomaly based on Vails temperature, even though their average temperatures will be significantly lower than Vails. These areas also have smaller diurnal (daily) differences between the high and low temperatures than the many other sites that are imputed into the average daily, monthly, and annual temperatures.

Additionally, data, from much of the lower elevation areas that makes Vails anomaly warmer than it truly is, exerts a strong warming influence that is imputed to these higher elevation areas. This adds up to a significant warming bias, but, wait, there is still another, even more sinister warming bias, which results from the "correction of raw data". That brings us to the second questionable methodology:

The logarithm that corrects the raw data is constantly scanning the data coming in from the various sites and comparing each data point to the sites nearest to it. If the data anomaly from a particular site shifts to a significantly cooler value than the surrounding sites, the data may be dropped, or adjusted back to the "pattern relationship". This is done because if data suddenly shifts from the "normal" relationship pattern, it must be the result of an error (a faulty assumption).

The vast majority of these detected "errors" are those which yield a cooling effect, thus requiring an upward temperature adjustment, as is manifest by the size of the required adjustment in the anomaly.

Still using Vail Colorado as our reference station let us look at how wrong these changes to the raw data are. Most springs and periodically during the winter, warm dry air blows down the eastern slopes of the Rocky Mountains resulting in sudden temperature increases that can be as much as 40 degrees F in a very short time. These systems may warm the western plains and the eastern face of the Rocky Mountains from Colorado to Montana, and at times as far north as Elberta Canada.

These weather events, called Chinook winds, change the temperature relationships between the large numbers of weather stations in the western plains and along the front range with minimum impact on temperatures of the weather in places like Vail.

These changes may amount of 20 to 40 degrees in a matter of minutes and may trigger the correction logarithm to "correct the data" and drop the cooler data observations or change them to "match the pattern relationship".

There is one reason why the data changes are most often in the direction that will result in a warming effect, it is bias. Statistically, with a vast number of data points, errors resulting from operational, equipment, human and miscellaneous errors should result in the warming errors canceling out the cooling errors. That does not occur in this system. Is that suspicious? The answer is an unequivocal, Yes.

The 0.25 degree C warming adjustment for "data based" anomaly errors is a huge adjustment. They are not warranted and they are indefensible. That figure does not include an estimate of the contribution of the "sea surface error" nor does it include an estimate of the altitude bias in the system.

Remember that, as you read the remainder of this work. Every time I give a figure of the total anomaly error estimate, it does not include an estimate of these two factors (altitude error and temperature spreading error). It would require a great deal of intense work to come up with a reasonable estimate of the warming effect of these "errors" which would require a downward adjustment. It would predictably result in a significant decrease in the current and past anomaly figures so it will not be done.

When you add the admitted +/-1^{0} F basic system wide error, to the urban heat island error, the sea surface temperature error, and this altitude related warming error you have an error factor that renders the system useless at best, and misleading, in the worst case scenario. Then when you add "data correction," "data set homogenization" LOWESS smoothing and long term rolling average distortions, to the error stew one must ask: What kind of a "Fraud"(*1) is this?

Yet it is so widely accepted without question? How is that? It is time to read and study Appendix #1, "The Normalization of Variance and Groupthink" p. 260. Many of the points found in that appendix apply to this situation.

Increasingly, the real questions is; Has this just happened, or was this pathway, to what is rapidly becoming a state sponsored secular religion, something planned and promoted? This question deserves deep contemplation.

The top graph on p. 200, was found in an article entitled "Thorough, not thoroughly fabricated: The truth about global temperature data" in an ars Technica publication @ (https://arstechnica.com/science/2016/01/thorough-not-thoroughly-fabricated-the-truth-about-global-temperature-data/). Look at it closely. It illustrates another fatal fail. Note: even though most of this article deals with the US data, the title is "The truth about global temperature data". This infers that, everybody does it (and they do, even NASA) so it is systemic throughout the entire GISS world wide system.

Every time you see this graph you should realize that the change to the 20 year base period of 1900 to 1920 accounts for approximately 0.3 degrees of the anomaly in the raw data (vs. use of the 30 year period of 1951 to 1980 base period). You should also remember the answer to the first question in the GISSTEMP section: "To measure SAT we have to agree on what it is and, as far as I know, no such standard has been suggested or generally adopted."

If you are free to change and shorten the base period when it changes the year to year global mean anomaly so it looks nothing like the 1951 to 1980 base period pattern; Have you defined a standard which has been generally adopted?

The top graph on p. 200 is the product of Berkeley Earth, one of the leading authorities on data sets, correcting errors in the recorded data, and homogenization of data sets coupled with smoothing. All of these should be regarded as non-scientific methods, but, because "everyone does it" they are accepted and used in many ways to "strengthen the case" for the global warming/climate change doctrine.

Viewed together, as shown below requires a reminder; these two graphs represent the temperature anomalies for the same period (1880 or 1900, to 2014) and for the same planet (just a reminder as it is not apparent from these plots). As you examine and compare the graphs below, [the one on top shows the US Raw Data line (red) the TOBs-Adjusted line (blue) and the

Homogenized line (green)], realize that only the red Raw Data line is real. The other two are recently "generated lines" based on altered raw data, processed data, generated (extrapolated) data and many data sets all mixed together in a "homogenized" stew like manner. Make note of the following:

The line graph from Berkeley Earth is based on a 1900 to 1920 base period. If you consult the 1880 to 2014 annual temperature anomaly chart it is evident why this 20 year period was used for the baseline. It is the only period since 1889 where the annual anomaly was equal to -0.46 degrees C. It is also curious why a 20 year time period was used instead of the usual 30 year, generally accepted length of a climatology.

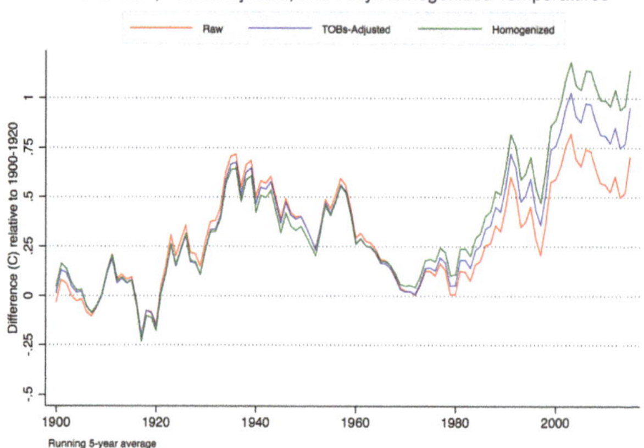

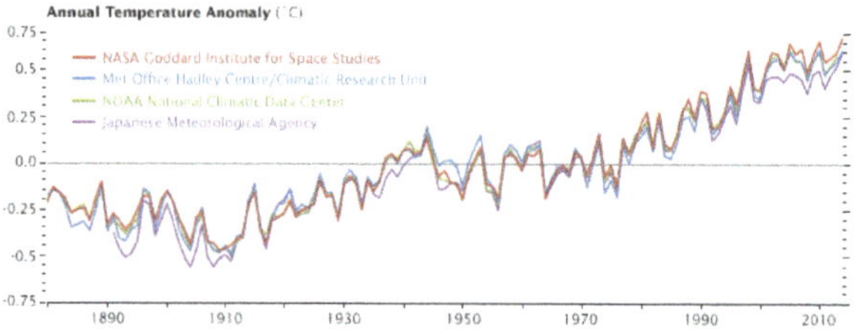

That is, until you look at the anomaly chart and see that inclusion of the 10 years previous or subsequent would have raised the baseline significantly. (This "base line" results in a significantly increased upward slope when any predictions of future temperature increases is made: a coincidence?) Yes, I am saying that there appears to be a little manipulation going on here. The change in the base period changes the entire pattern of the global mean anomaly (That is why I remind you; these charts cover the same time period on the same planet).

1. Note how all three lines in the Berkeley Earth graph are highly consistent up to approximately 1970. The people taking the raw data were doing a terrific job during that period, the equipment was functioning well and everything was going great. It should be noted that beginning with GHCNv3 version of the data set a significant number of corrections to the earlier data have been made. These changes have increased the annual global anomaly slightly but have increased the predicted 100 year change in the anomaly significantly.

2. Note (in the top graph) the huge variation in the data between 1951 and 1980 which serves as the base for the world-wide GISS Temp system. The anomaly dips down under -0.25^{0}, then increases to over 0.5^{0}, then drops to zero in the mid 1970s and again in 1980. The wide swings in the anomaly between 1951 and 1980 means that the claim that this was a period of relatively stable surface temperatures is not correct. The fact is, the 1951 to 1980 time period as shown in the two anomaly charts above are vastly different. Which

one is correct, or are they both wrong? Rhetorical Question: Remembering that one of the stated reasons for developing the GISS anomaly system was "What exactly do we mean by SAT? Answer: I doubt that there is a general agreement how to answer this question". Now that we have GISS SAT do we see general agreement?

3. One thing that is readily apparent when you compare the above charts is that the base period chosen makes a huge difference in the relative and absolute values of the anomalies calculated from them. This illustrates that the system can be easily manipulated to obtain the results desired. Is that a scientifically defensible quality, or an indication that the entire system is readily adapted to accomplish fraudulent goals and/or purposes?

4. Another notable incongruent feature of these charts is the huge anomaly that was reached in the late 1930s (according to the top chart). The anomaly increased to approximately 0.7^{0}C, a level not seen again until about 2004. It should be noted that the atmospheric carbon dioxide concentration at that time (1930s) was only 300 ppm, slightly increased (by 20 ppm) from the pre-industrial level. So what caused this marked warming? It was not carbon dioxide.

5. Now concentrate on the 1980 to "current" part of the top chart. Notice that even though there was a rapid rise in the anomaly (equal to approximately 0.6 degrees for the raw data, of which 0.3 degrees is attributable to the base period chosen) the lines remained fairly close until

approximately the year 2000. Apparently the people and equipment were still operating very well, up to 1980 and quite well up to 2000 because not much correction was needed and the homogenization effect was not great. Also, remember that 1980 (part of a very cold period in the northern hemisphere) is used as the baseline for many global warming indicators such as Arctic Ice Extent.

6. Finally, examine the two charts closely during the period of 1998 to 2014. The global anomaly is nearly flat, while the Berkley Earth chart shows the anomaly increasing by over 0.6^{0}C for the corrected and homogenized values and approximately the same for the raw data. This chart is based on a 5 year rolling average which should have the effect of dampening sudden changes, yet there is no indication that occurred. Can you spell "Suspicious"? That would negate most of the post 1998 anomaly increase in the Raw Data (that is very significant).

Authors note: The correlation between the change in the annual anomaly and the increasing CO_2 concentration is actually negative during the period of 1880 and 1920. In other words, as the CO_2 increased the annual global anomaly went down. Was the increasing CO_2 content of the atmosphere causing the earth's surface temperature to cool? This illustrates a number of internal inconsistencies of the theory and the outlandish assumptions it is based on.

Why do the three lines begin to increasingly diverge (separate), beginning in 1980, then become even more divergent after 2000? It is confusing with significant internal inconsistencies. Here are the reasons given in the above referenced article (and in many other publications), as to why the data corrections and homogenization is necessary.

The adjustments were justified on the basis of several error factors which included:

1. Much of the data was taken and recorded by volunteers,
2. Some observation station locations changed,
3. The time of day that data was read and recorded was changed, and
4. Equipment changes were made from recording Hi-Low thermometers, to electronic thermometers, then to modern data recorders. ({cf} Satellite observations of world wide surface temperature done through MODIS, indicates that the raw data is quite accurate, if the 0.3^0 base period error is removed). I will briefly deal here with these four issues as it pertains to the Raw Data and TOBs-adjusted data. I will deal with the homogenization issue later in more detail.

Issue 1: Volunteers and the errors they make: Look closely at the above chart and ask yourself: what caused a very reliable group of thousands of volunteers across the US to suddenly become increasingly careless and error prone beginning in the late 1970s, and even more so after 2000? They did an excellent job before that point in time, then an "error correcting algorithm" was added to the equation. <u>Rhetorical Question:</u> Did thousands of volunteers, NOAA employees, and administrators suddenly become careless and incompatent, or does the algorithm reflect the biases of its authors?

This graph proves the specious nature of this allegation and calls into question the use of the correcting algorithm to change the raw data. It appears to be an attempt to fraudulently obtain a desired result.

Since 1980 more and more of the NWS sites (and weather stations world wide) are equipped with modern electronic data acquisition systems, which are extremely accurate, precise, durable and well maintained. And yet the divergence of the three lines increases from less than 0.1^{0}C for the period 1950 - 1976, to 0.15^0 by 1980 and then increased to 0.45^0 by 2016. (note dates are approximate-read from graph axis).

The time period, 2000 to 2017 involved a time when almost all the data was collected by modern, highly accurate instrumentation, making the claim of increasing errors in that period, specious at best and an attempt to perpetrate a fraud at worst.

Issue #2: Some observation station locations changed. Is it a believable assertion that when station locations were changed, a preponderance of them were located in less representative (cold) sites, in terms of obtaining accurate data? That is the implicit assertion. How non-professional and incompetent those in charge of the locations must be (Do you believe that?).

This practice, of adjusting the temperatures if the new station location yields data that produce a different anomaly than the previous location, <u>exposes a basic fatal fail in the GISS scheme.</u> Remember the 1200 km radius circles we discussed before? The system authors maintain that the temperature anomaly within that radius is uniform. A flawed assumption, as indicated by a need to make these adjustments (another internal inconsistency).

I have known and worked closely with a number of NWS professionals over the last 45 years and have found them to be very professional and competent. In addition, as times and technology changed and improved, the modern weather stations became more and more accurate; not less and less as stated, and inferred by the above article and the accompanying graph (Do not forget who and what this graph represents, can you say, Changed Raw Data, Homogenization, and Smoothing.) Two of the three lines on this plot are fabricated, not scientifically generated. The third line, the Raw Data, has a 0.3 degree increase in the anomaly which is traceable to the use of the 1900 to 1920 period as the "normal temperature". The author realizes the last sentence is redundant, but the extreme importance of this point justifies redundancy.

Very important Rhetorical Questions:

1. If the huge investments that have been made in modern equipment are resulting in greater required adjustments in terms of frequency and/or magnitude of errors, what justifies the investment?

2. If, as maintained in the above article, the satellite based systems of surface temperature monitoring are not as reliable as fully "corrected" and "homogenized" earth surface station data sets, what justifies the maintenance and expanded missions of these very expensive systems. (Update: It appears that the data flow from these systems are beginning to be used in the evolution of the anomaly in a different way: See: Evolution of the anomaly at: <fatalflaws.net>).

3. Do you realize that the changing of the raw data the way it is done decreases the calculated total error factor significantly (it separates it from reality)? Could that be

the reason that it is done? Would that be Fraud?

4. A Statement: As I read through the critiques contained in this article it reminded me of someone who believes in settled science, letting their bias run free in search of a justification for their position. Modern spectroradiometry technology does not resemble what the "authorities" quoted describe (it is close to 1975 technology). Current technology is able to reach through the atmosphere to the earths surface with very narrow and specific bandwidth settings, accurate enough to distinguish between emissions given off by carbon dioxide, nitrous oxide and methane. They can also do things such as map algae concentrations in the seas by their chlorophyl content. This is another example of the limits and bias of "Settled Science". (Refer to Appendix 1: Symptom #'s: 1,2,3,4,5, and 8; p. 260: You be the judge)

Issue 3: The time of day that data was read and recorded was changed. It is true that NWS changed the time of day temperatures were read and recorded, from evening to morning and the change took place gradually for many reasons. One of these reasons was the adoption of daylight savings time in the early 1960s. The change was made largely to avoid the alleged error described below.

Allegedly, the biggest error factor cited by global warming advocates to justify numerical corrections, or dropping data that is deemed to be erroneous is that; If the high temperature was read too early, the recording thermometer reset, and the temperature the following day was cooler, the "recorded temperature" would be too high for the subsequent day (this would result in a warming bias and require a cooling not a warming adjustment). The changes which are made are preponderantly corrections of a cooling bias (revised data is warmer), as proven by the TOB's (blue) line being warmer than the raw data.

To be specific: If a high/low thermometer is read, the data recorded and the indicator reset, before the high temperature for the day is reached, the indicator will move with the liquid to the actual daily maximum and stay there. For example if it was reset when the temperature was 65 degrees and the temperature increased to 75 degrees, the next day the reading, of that days high, would be 75 degrees, even if the actual temperature had not reached 60. This is what is described and if there were a true correction made, it would be to lower the temperature, not raise it. (This is an example of why true science and data correction is incompatible - - it opens the bias gates.)

The time of day change would tend to minimize this problem, which was not a problem prior to the late 1970s (check the plot) but became a problem of increasing magnitude between then to now. This is nonsense, It should require adjustments downward, to compensate for the warming bias, not upward which is what the graph shows happens in most cases.

According to global warming dogma the recording of the overnight low in the morning introduced another cooling error. Is this believable?

Issue 4: Equipment changes were made from recording Hi-Low thermometers, to electronic thermometers, then to modern data recorders and complete automated data acquisition systems, even Satellite based systems such as MODIS. The claim is that the improved equipment led to a significant cooling bias that requires (warming) correction. Are these credible claims? I am sorry if that last question was taken as an insult to your intelligence.

I have worked with data recording equipment for 60 years and the improvements have been incredible. Since the 1980s I routinely use raw State Climate Center data (from three different states as well as NWS data).

This data is collected at 10 or 15-minute intervals, is extremely accurate, and is very dependable. It is used to run disease and insect developmental models to determine the need for, and timing of, control measures.

If there was a cooling bias in any of this data it would result in failure of the insect and disease integrated pest management systems used widely in modern agriculture in the US and most other countries. I have not experienced that in the last 38 years.

Remember, according to the above graph, the period of increasing errors in data, corresponds to the period when increasing use of automatic, very reliable, and accurate, data acquisition systems became the norm. This makes the divergence of the corrected data and homogenized data lines from the RAW Data in the above graph totally absurd. It is contrary to my experience and the experience of everyone I know who deals with and depends on accurate temperature data.

If you believe this Global Warming Dogma please contact me: I have some "ocean front, beach property" on the San Rafael Swell in central Utah for sale. I guarantee that it was once ocean front, or ocean bottom, and the fossil record proves it.

Bottom Line: It is ludicrous to claim that the quality of temperature data has declined (hence needing increasing levels of warming adjustment) from the mid 1970s to the current time. Essentially this is the claim. The equipment is increasingly accurate and reliable, so the claim is subtly asserting that the skill and professionalism of those using it to collect data is deteriorating. Another specious claim to justify raw data manipulation and/or manufacture: That is not science, it is witch-craft or worse.

Realizing that they (the two graphs above) are both considered reputable and accurate sources ask yourselves: How much confidence do I have in either one of these highly divergent plots, or the people in charge, and the systems themselves, that produce this unbelievable information?

Now let us move on, to the third methodology that raises questions about the credibility of what we see and hear so frequently.

To Set The Stage: An Allegory:

<u>The "Freshly Pressed, Ice Cold, Apple Cider Allegory"</u>

A home-owner who was into gardening and raising apples came to apple harvest time in the fall with an abundance of fruit on his trees. He was an advocate of "Organic Gardening" who had not began to use mating disruption, so many of the fruit had at least one worm hole in it. He picked the fruit and as he filled the four baskets, he separated the best fruit into one basket (those with no worm holes), the wormy fruit (those with 1 or more worm holes) into another basket, the blemished fruit into another and the soft fruit into a fourth. When he finished he looked at the baskets of fruit which, were heaped with apples and considered each.

Basket 1: Contained the best fruit. He was proud of it, large, crisp, juicy, excellent fruit.

Basket 2: This fruit was fruit that had one or more worm holes in them. He knew that all the worms had matured, left the fruit, found a place to pupate over the winter and complete their life cycle the next season. All that was left of those pesky things were the holes, and yes the apple tissue that had passed through their gut. He had heard a professional entomologist refer to this as "frass," a nice scholarly term, much better than worm poop.

Basket 3: This fruit was very nice fruit, but had one or more blemishes on them. They were just right for cider, after all the blemishes were only skin deep. And the worms loved them also (as indicated by the frequent worm holes).

Basket 4: This fruit was large, had been crisp, was very juicy, just that it was a little like the apples you sometimes buy in August that have been in storage for 12 months. It was not rotting, but would be soon if he did not use it.

He arranged with a friend to borrow a cider press in two weeks to juice his apples, and not having room elsewhere he sat the four baskets under a large shade tree. He watched the baskets closely but noticed each evening a large dog and two smaller dogs pass the baskets, lift a rear leg next to baskets two and four and mark their territory. He thought, I should stop that, but, after a long days work I deserve to relax with my coffee and news program.

On the day he had chosen to press the apples he had the thought: I really want 10 gallons of apple cider, and I know that each basket will yield about 2.5 gallons. If I use all the fruit I can get the 10 gallons, and besides there are no worms in the fruit with worm holes in them, only the frass and possibly a little mold in the worm holes. The soft fruit seemed to be very juicy, and the blemishes were only an appearance issue. Each bushel weighed a total of almost 50 pounds and after the dog urine had evaporated there could not be more than a few grams of residue, so what is a few grams out of 90,800 grams of total weight (he was conversant with the metric system).

So, he juiced all four baskets and ended up with 10 gallons of cider. Oh, congratulations, he has invited you and your significant other over for a night of cards and fresh pressed, cold, apple cider and doughnuts.

I'll let you consider what this has to do with, anomalies, temperature spreading, removing "mistakes" from raw weather data, and homogenization of climatic data sets. After finishing this chapter pause and consider how this allegory applies. You are not invited over for fresh, cold apple cider, but you are invited to read on and consider what you read.

You may want to print out Appendix #1 that deals with "Normalization of Deviance and GroupThink" (p. 260) and keep it handy for quick reference. See if you can recognize any of Vaughn's eight symptoms, or the ones I added at the end.

Homogenization of Data Sets: To avoid another 100 pages, this discussion will be very general. (Excuse me for asking you to make three assumptions in close order.) 1. Assume that you have an "moderately" accurate data set that describes the global surface temperature now and for the past 140 years. 2. Further assume that this data set has a defined reasonable error range for the purpose for which you intend to use it, and 3. Assume that you have 13 other data sets, each with known and/or suspected significant flaws.

Rhetorical Question 1: Do you use the most accurate data set, or do you mix them all together (homogenize them) so they are all reasonably suspect but the bad ones are a little better?

Rhetorical Question 2: If you have "settled the science" isn't there a requirement that you have a reliable data set that doesn't need to be changed frequently?

Rhetorical Question 3: Remember the statement contained in the answer to the third question we discussed from the GISS SAT document that said: "For the global mean, the most trusted models produce a value of roughly 14°C, i.e., 57.2°F, but It may easily be anywhere between 56 and 58°F. . ." Does the phrase "the most trusted models produce a value of roughly 14°C" instill confidence in your mind concerning this system? The phrase "roughly 14°C could be anywhere from 13.33 to 14.44°C. What do the other (less trusted) models say? Are they in whole or in part, the other 13 datasets?

Rhetorical Question 4-6: Do you wonder what the parameters are that determines "the most trusted models"? Do you wonder why there are trusted and untrusted models? If the science is settled, shouldn't there be a single, definitive, trusted model using the "settled dataset" instead of multiple "most trusted models," and yes a non-specified number of untrusted models? I could go on and on, but you get my drift. Unfortunately the disciples of global warming probably never will, not after three decades of investment in a failed theory.

The answers to the first two questions above are obvious, but what if you do not have an accurate data set as described by 1 above? (By accurate you mean one that gives you the desired answer.) The answer appears to be (within the settled science crowd of the global warming disciple community) mix them all together and hope the result gives you what you want; If that doesn't work, choose the mix that does.

Allow me a little digression that bears on the above topic and what is to follow: (Please; the magic word I try to teach my grandchildren).

I have hesitated to address the following subject but because of its impact on generally accepted methodology, including being adopted by the datasets used by NASA (and other global agencies) beginning in 2011 it is necessary. The subject is what has been called "Climate Gate" which began circa 2009.

An external hacker obtained access to a server at the Climate Research Unit at the University of East Anglia (UEA) and "hacked" a large number of emails and other documents which were then published.

This led to a heated controversy between the "disciples" and "deniers" of global warming with one of the points of controversy being a question asked by one advocate; "how do we hide the dip"? This was in reference to a drop in global temperatures (between 1998 and 2009) while the CO_2 concentration was rising. The answer was; Use the "Mann trick," in a reference to a methodology employed by Dr. Mann the author of the hockey stick theory (Dr. Mann used the term trick in one of his emails to describe the method).

In simple terms the technique is to use homogenization of a different data set (or a portion of one) with the one the researcher has been using to get a positive outcome (an outcome that is more doctrinally friendly or more striking i.e. in this case one that would reverse the downward trend of the anomaly).

Dr. Mann claimed that the use of the word "trick" was taken out of context and that the technique was a recognized procedure often used in climatic research. There did not appear to be a denial that the technique changed the result. Rhetorical Question: If a procedure or technique is used for the purpose of changing an experimental result is it "a trick"? (I am sorry, if you took that question as an affront to your intelligence.)

A Penn State panel examined whether Dr. Mann "seriously deviated from accepted practices within the academic community for proposing, conducting, or reporting research or other scholarly activities." The panel concluded that he had not.

Dr. Mann and the disciples of global warming were elated and claimed vindication. Rhetorical Question: I described briefly what the procedure is. Does the fact that it did not "seriously deviate(d) from accepted practices within the academic community" serve as vindication, or, as a much broader indictment?

This question is extremely important because as you will see in the remainder of this chapter, the changing of data, the homogenization of data sets, use of rolling averages and use of LOWESS smoothing often change the results and/or unjustifiably accentuate trends.

Rhetorical Question: Does the fact that these practices do not "seriously deviate(d) from accepted practices within the academic community, for proposing, conducting, or reporting research or other scholarly activities" give you confidence in the results? True science is not a "everybody does it so it is ok" proposition.

If you closely examine the graph below, showing the raw data, corrected data and homogenized data, you will notice that from approximately 2001 through 2009 the trend of all three lines was downward from 0.85^{0} C to 0.5^{0}C for the Raw Data (which would be 0.2^{0}C if the bias of the different base period was removed). The adjusted and homogenized lines followed the same trend, then all three rapidly increased following 2011.

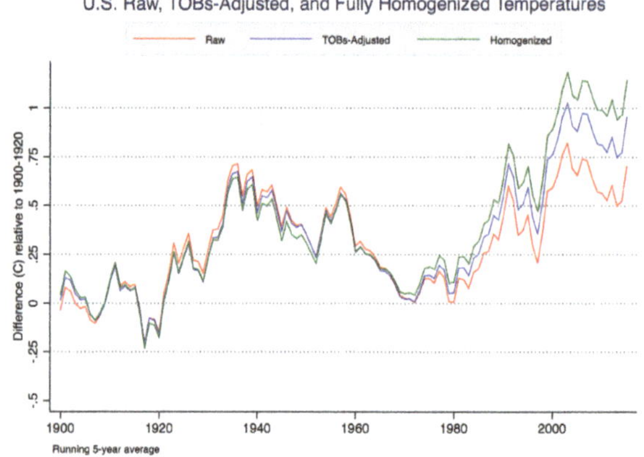

<u>Rhetorical Question:</u> Why did you start driving your SUV again after eight years of blissful existence with decreasing temperatures and atmospheric carbon dioxide?

The answer to the above rhetorical question is found in the materials that brought about "climate gate," "a trick of the trade" was employed to "hide the dip". As it turned out the different investigations into this matter found no wrong doing on the part of those who were involved because the things they did were commonly used methods. <u>Take a moment and consider that fact.</u>

That was not vindication, but another level of indictment of the commonly used methods. In effect you generate new numbers by throwing in another data set, or additional data and homogenizing it all together. I realize the preceding is repetitive; but given the importance of this topic it is worth it (at least in my opinion).

<u>Rhetorical question:</u> How many gallons of "frass-free" apple cider could have been had, if none of the worm poop and dog pee was allowed to contaminate the apple cider, in the allegory at the beginning of this chapter? I listened to a radio talk show host a few years back that often said: "It doesn't take much urine to ruin a delicious malt."

This is, to my way of thinking, and training, a totally non-scientific, even anti-scientific procedure. Simply stated this trick allows a researcher too; when a researcher fails to obtain the result he wants when he "runs his model" he uses a different data set, hybrids it with his own and re-runs the model. If he chooses the right data set, and their vagrancies, deficiencies and tendencies are well known, the non-significant result is, presto, turned into a significant one.

The committee of Dr. Mann's peers who investigated his "hacked emails" that discussed using this trick as a method of "hiding the dip" found that his trick (of the trade) was widely used and hence was not a transgression. May I observe that, using a method such as this is much easier than employing true science, because it guarantees that you don't waste effort arriving at unpublishable results.

It also results in one more "peer reviewed" publication to grease the skids toward tenure, or other promotions (to say nothing about how it supports a fatally failing theory).

Here is another example of how questionable practices are used by disciples of global warming/climate change to obtain the results they want. Quoting from one of the questions and answers contained in the Global Surface Air Temperature article we discussed earlier: Q. "*If SATs cannot be measured, how are SAT maps created?*"

"This can only be done with the help of computer models, the same models that are used to create the daily weather forecasts. We may start out the model with the few observed data that are available and fill in the rest with guesses (also called extrapolations) and then let the model run long enough so that the initial guesses no longer matter, but not too long in order to avoid that the inaccuracies of the model become relevant."

The first part of the above answer which reads; This can only be done with the help of computer models, the same models that are used to create the daily weather forecasts. We may start out the model with the few observed data that are available and fill in the rest with guesses (also called extrapolations)...." There are two important points raised by the last sentence fragment quoted above. They involve:

First. "the few observed data that are available". In any scientific analysis, the fewer the data points available the less confidence can be placed in any conclusions drawn.

Second. "And fill in the rest with guesses (also called extrapolations)...." What do guesses and extrapolations add to the answers or conclusions? Two things; doubt and the potential for ***bias***, nothing more. Rhetorical Question 1: When one is dealing with what they claim is settled science, and it is easily knowable that this is not the case, can one make independent, intelligent guesses, or extrapolations? Rhetorical Question 2: Does not the claim of settled science, preclude consideration of any guesses or extrapolations which do not support, the unfounded notion of the science being settled?

Authors note:The above answer falls far short of meeting any scientific quality standard. The sentence fragment above that reads: "We may start out the model with the few observed data that are available and fill in the rest with guesses" is surprising. The data that are available runs into the tens of millions of temperature readings from many sources world wide. With that mass of data, how can "filling in the rest with guesses" add anything? If there is any bias in the guesses, the simple answer is that the guesses change the result toward the bias directed goal.

The last part of the above answer which reads: "then let the model run long enough so that the initial guesses no longer matter, but not too long in order to avoid that the inaccuracies of the model become relevant.. . ." is even more sinister and problematic. The first portion of the above sentence fragment appears to be a statement that by letting the model run (make sequential runs using the same data and information) the deficiencies of the guesses and extrapolations can be corrected.

If the guesses and extrapolations are not actual additional data, this repeating of the analysis cycle would actually make them worse, not better. Now follow along because here is where the procedure becomes sinister, i.e., predictably mis-leading.

If one makes the assumption that each run of the data is a replication of the "experiment," then the probability that a given numerical result will be significant is increased by each sequential run. This procedure would be the equivalent of a scientist copying his experimental results over and over and considering each copy a repeating of the experiment, which it obviously is not.

It is an anti-science trick to do this, and should not be countenanced. In my time as a graduate student, researcher, or graduate advisor that behavior would have been regarded as so dishonest that it would bring to an end the career, or potential career of the one using the technique.

Then comes the last phrase: "but not too long in order to avoid that the inaccuracies of the model become relevant. . ." Each "replication" of an experiment carries with it an error factor, which is additive to the errors of preceding and following replications. If these errors are too large the results cannot be judged to be significant. What the above is saying is, you gotta pull the plug on the computer before the inaccuracies of the model (and data, real and guessed) become relevant, i.e., give you the true result (non-significance).

With the above in mind consider the following Rhetorical Question: When you see the above examples of what is called "climate science" do you understand why my judgments of the theory and the people behind it is harsh and judgmental?

Do you agree? Spend a little time looking at the graph on p. 216 and consider how much intellectual honesty you can see as you observe what is patently false science. Then observe how much your intellectual temperature rises.

What ever your thoughts about this Rhetorical Question are, please read on.

Photo below: Earlier, I included a photo of a dangerous portion of the trail to hidden canyon in Zion National Park in southern Utah (USA). The photo below was taken in the upper reaches of that canyon. To get to this point you must traverse the segment of the trail shown on page 5 with the chain fastened to the canyon wall. I wrote about that segment of the trail ; "To cease to "look" and examine what we think we know for bias, is equivalent to moving along this trail with our eyes closed and not holding onto the chain." The 500' drop from the trail has proven fatal to a number of people over the years who did not keep a good hold on the chain.

Contemplate what effort this area might warrant, if all that is needed to get here, is to hold onto the chain of objectivity and constantly examine our thought and belief patterns for bias. It is nature at its best. Beautiful, cool, and quiet, even when the temperatures outside this canyon are 110ºF+, and there are 25,000 visitors in the park. I have seen birds, lizards, and mice footprints, bighorn sheep and a mountain lion in this area, but not a single human.

For more on these topics visit: <fatalflaws.net> Pay particular attention to "Evolution of the Global Anomaly"

Chapter 11: Summary: Who and What Can be Trusted?

<u>Synopsis:</u> *We have discussed and established how the following failures of the Global Warming/ Climate Change doctrine have resulted in a theory that, on its face, disproves itself. These six failures (Fatal Fails) are indisputable.*

1. The failure of the Global Warming theory to include the effects of; the sun's role in determining the earth's surface temperature.

2. The failure of the Global Warming theory to include the effects of; water's role in controlling and modulating the earth's surface temperature and climate.

3. The failure of the Global Warming theory to include the effects of; the role of heat from the earth's core in influencing the earth's surface temperature and polar ice caps.

4. The failure of the Global Warming theory to include the effects of; the role of atmospheric particulate in the greenhouse effect.

5. The failure of the Global Warming theory to include the effects of; the important roles played by, and the fate of, atmospheric carbon dioxide. It is not a permanent addition to the atmosphere that is not subject to temperature dependent physical and chemical changes, so it is not a climate forcing agent (by definition).

6. How the advocates of the Global Warming doctrine have failed to recognize the effects of: 1. bias, 2. mis-use and abuse of data sets and 3. changing the numbers; on the credibility of the Doctrine and the Disciples of this theory. It is a failed theory, not established science, because it was and is based on an erroneous assumption and ignores the interactions that are known to be taking place between the active, climate controlling forces.

Chapter 11: For your contemplation I present below five graphs. Examine them closely for differences and inconsistencies then contemplate the questions that follow.

The first graph below was copied from a Washington Post article published 4/21/2019 entitled: "Earth Day affirmation: New satellite data confirms Earth is warming and possibly at a faster pace." This article focused on the Arctic and ice loss. There was no mention of the role of natural variation in the melt of 40% to 50% of the Arctic ice melt over the last 38 years due to the prominent ElNino, even though it is indicated on the graph. There was also no mention of the ice melt that was due to the ozone depletion of the upper atmosphere and its impact on the arctic ice melt. (when you are an advocate of a theory that is flawed, you don't want to point out the theories weaknesses; **advocacy is an antagonist of science**). The graph is a NASA product.

The graph below is one you have not seen before (in this work). It is very instructive. Note how all five lines follow the same pattern from 2004 to 2014 showing a negative anomaly before 2014 when the anomaly became positive. The anomaly reached +0.25 to +0.35 by 2016 (variation between sources). It will be important to compare this anomaly with the other graphs shown below.

These are some of the most credible sources, and the AIRS (red line) is, as indicated, from the Infrared satellite data. The technology used in acquiring, recording and presenting this data is the "Gold Standard" unless the data and function is fudged. (See:Evolution of the anomaly at: <fatalflaws.net>) As indicated, the baseline (0.0) is the global mean surface temperature for the years 2003 to 2017.

Note 4 Things:

1. This plot again raises the question of what is the standard or normal. A casual look can be very misleading if the 0.0 base line is assumed to be the "standard" 1951 to 1980 "Normal". This "Normal" is the mean for the period of 2003 to 2017. <u>Rhetorical Question:</u> I have asked this question before: Was not one of the justifications for the GISSTEMP SAT system that there was no generally accepted method of determining the SAT?

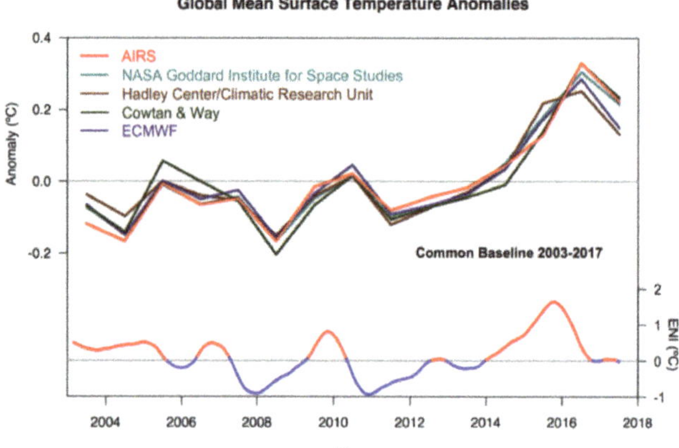

NASA AIRS Infrared satellite climate data, and El Nino cycles

2. Note the decrease between 2016 and 2017. If you consult the interactive annual global mean surface temperature anomaly chart at: <https://climate.nasa.gov/vital-signs/global-temperature/> and hover over the 2017 value and the 2018 value you will see that in those two years the anomaly dropped from 1.02^0 (2016)*1 to 0.9^0 and then to 0.82^0 for a total decrease of 0.20^0 C. This decrease is equal to 19.6%*1 of the total historical anomaly

increase (based on the 1951 to 1980 baseline). Rhetorical Question: How does this happen for two consecutive years when the CO_2 concentration was rising at record rates? Could it be that the theory is fatally flawed, resulting in fatal fails? *Authors Note: Notice the rapid increase in the ENI anomaly between 2014 and 2016. During this period the ElNino conditions north of the equator in the Visayan region of the Philippines was particularly strong. The strong ascendant forces of rapidly evaporating ocean water resulted in severe drought conditions across the entire Philippines. In four years normally this area would produce eight good rice crops. In that period the drought was so intense that two partial crops were all that survived. This is consistent with the Lawrence Livermore Laboratories article on the relationship of ElNino and unusual warming of the Arctic. The ElNino cycle shifted to a prominent LaNino in 2017. The rains returned to the southern Philippines and the anomaly began to fall in the Arctic, requiring a new data set be introduced in early 2019 (Part of my family lives and farms on Bohol Island in the Visayan region; read on).*

3. At the above interactive site, the plot shows the LOWESS smoothed value as the dark line. Even though there have been two years and a 19.3%*1 decrease in the anomaly, the LOWESS value is still increasing through 2018. This deception is evident on its face, but it will get worse if the 2019 anomaly decreases again (a very probable result). How does it get worse? If there is a decrease of 0.08^0 C or less, the LOWESS value will continue to rise with nearly the same slope as the last two years. Rhetorical Question: Is this evidence that I was correct when I questioned the use of LOWESS as an indefensible technique? I should rest my case on this one, but you will see an additional indisputable example later.

4. Since I first drafted this section, the GHCNv4 data set has been adopted and once again the

cooling trend has been reversed. <https://www.climate.gov/news-features/understanding-climate/climate-change-global-temperature>. At this website the 2019 value is 0.98^{0}C, while the actual number, even with the changes in the data has been reported to be between 0.90^{0} to 0.98^{0}C[*1]. These numbers resulted from the application of LOWESS smoothing (five year rolling average) and the other data manipulation procedures used in the v4 data set. (I personally prefer my fudge to be made from the traditional recipe which uses cooked chocolate, milk, confectioners sugar, and walnuts, not smoothing, and data manipulation.)

5. The El Nino/La Nino plot {labeled ENI(^{0}C)} shows how the El Nino activity correlates with the anomaly; Lending significant credence to the Lawrence Livermore National Laboratories Nov. 5, 2018 article I mentioned several times. It (this plot and the LLNL information) proves it is not all carbon dioxide climate forcing. Many have maintained that the 14 to 15 year pause in the increase in the earth's surface temperature did not take place, or was insignificant, an indefensible position, unless one accepts the multi-dipped fudged numbers used in the graph shown at the above link. You may wish to compare those numbers to the "raw data" numbers shown as the red line in the Berkeley Earth graph (p. 216).

6. Quoting the fourth bullet point to the left of the above referenced graph: "Despite a late-year La Niña event that cooled a wide swath of the tropical Pacific Ocean, 2020 came just 0.04° Fahrenheit (0.02°Celsius) shy of tying 2016 for warmest year on record." (At the interactive graph link in 2 above you will find that the 2021 anomaly had dropped again to 0.85; Will we see GHCNv5 in the near future?) The

above quote is an admission that the ElNino/ LaNino cycle plays a major role in the warming and cooling of earth's surface. This is not a human caused controlling mechanism. It has nothing to do with changes in the CO_2 concentration of the atmosphere (ElNino was dominant for the 38 years preceding this LaNino event). Don't count on global warming advocates to point out the obvious when it contradicts their biases.

7. This plot from 2003 to 2014 covers the last 11 years of the 14 year period when there was a pause in the increasing temperature of earth's surface mentioned in 5 above. There are five things this graph illustrates that are highly significant:

    1. In 2011 the GHCNv3 dataset was adopted which resulted in additional warming bias being interjected into the anomaly calculation. In the next 2.5 years v3.1 and v3.2 were sequentially adopted that added additional warming bias (you can see the upward inflections of the lines). For those reasons my analysis here will cover the years 1998 to 2011 (14 years).

    2. The 1998 anomaly was equal to 0.61^0C and the 2011 anomaly was 0.61^0C (based on NASA web archive data downloaded Sep 15 {2019} at 8:57 PM) The annual average mean global surface temperature anomaly for this 14 year period was 0.59^0C.

    3. In this 14 year time period the CO_2 concentration increased steadily from 366.27 in 1998 to 391.15 in 2011. How does that happen if CO_2 is such a powerful greenhouse agent? The CO_2

concentration increased by 24.88 ppm (equal to 21% of the 1880 to 2018 increase) and the anomaly remained essentially flat according to the unadjusted data. Simply stated: There is no correlation between the change in the CO_2 concentration and the change in the global average mean surface temperature anomaly during this 14 year time period. If the raw data was used the correlation would have been negative (as shown by the raw data line in the Berkeley Earth graph last seen on p. 216 of this work).

4. The AIRS anomaly increased by 0.3^0C between early 2014 and mid 2016. This rapid increase, equal to approximately 1/3 of the total anomaly increase since 1880, lagged slightly behind the very strong El Nino which peaked in late 2015, but remained at a significant level through 2016. (Remember, El Nino is a strong warming force, particularly in the Arctic and is not the result of human activity.)

5. As we discussed in Chapter 1 the solar minimum was approached in 2017 and through Feb. of 2021 the minimum remained the deepest minimum since the space age began. The CO_2 concentration in the atmosphere has continued to rise at unprecedented rates, largely due to increasing forest loss due to wildfires as discussed in Chapter 7, but the global mean anomaly decreased during 2017 & 2018 by 16.3%. With the adoption of GHCNv4 the anomaly increased to 0.96^0C (since changed to 1.02^0C)*1 by 2020 then

dropped back to 0.85^{0}C for 2021. This is not characteristic of a correlation between CO_2 and the anomaly.

The items 1 through 5 above is why the adoption of the GHCNv3 in 2011 and GHCNv4 in 2019 had to be made. Once you begin fudging the data there will probably be more and more fudge required. It's kind of like the real fudge, once you start eating it, there is no stopping.

The facts and figures presented above are an example of how multiple variables interact to modulate and control the earth's surface temperature. If CO_2 was the only factor, the temperature anomaly should have steadily increased during this entire period. It did not increase from 1998 to 2011, then increased rapidly for 5 years, the result of data manipulation. The mean global temperature anomaly decreased significantly in 2017 and 2018 (a trend that was reversed in 2019 with the introduction and use of the GHCNv4 data set). All this occurred during a period when CO_2 in the atmosphere increased at record rates. <u>(As a reminder: in both the GHCNv3 and GHCNv4 datasets the number of temperature measurement sites increased significantly, to over 13,000 and then to 28,000+. These sites were homogenized into the data set used, and both times, a downward trend of the real data was reversed.)</u> Can you spell "climate gate"?

Now compare the GISS Annual Temperature Anomaly graph (Top graph below) with one you have seen before, The "corrected" and "homogenized" Berkeley Earth graph. In the lower graph the red line is from the GISS system, using a different base period (1900 to 1920) with minimal "corrections" (Raw Data). The "corrected and homogenized" anomaly (blue and green lines) between 1998 and 2014 increased much more than the raw data supports. The data presented in this graph (the second below) has a correlation coefficient that approaches zero when one considers the rather steadily increasing CO_2 concentration between 1900 and 2018.

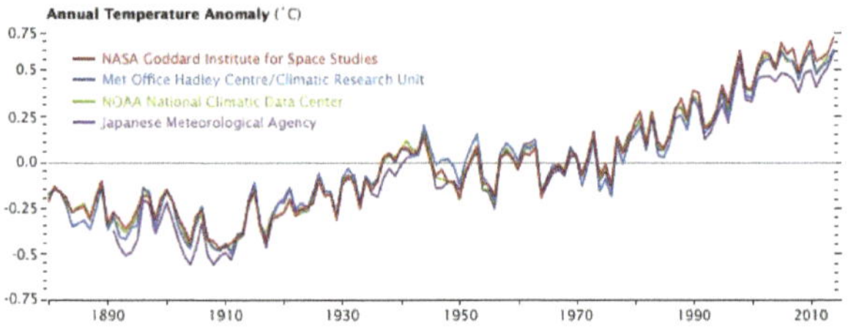

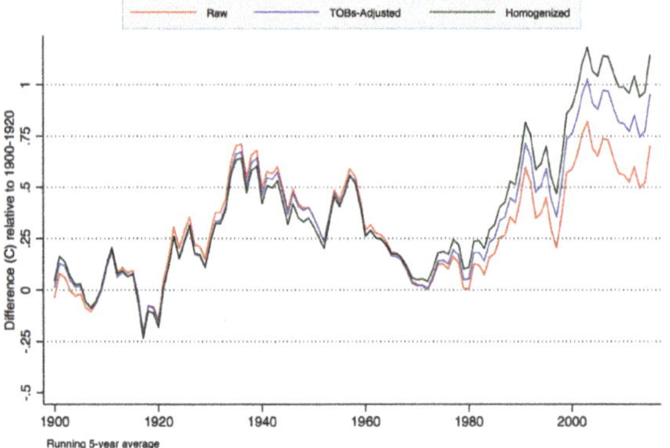

Rhetorical Questions:

1. Of what value is a system that can yield the vastly different results covering the same historical period?
2. What does it say about the global warming theory when the first graph above has a very weak correlation with the rising atmospheric CO_2 concentration and the second one has no correlation?
3. What do these two graphs say about the state of "climate science"? (See Evolution of the Anomaly @ <fatalflaws.net>)

4. Do you see the lack of internal consistency that is pictorially illustrated by these plots?

The graphs should look at least similar. After all they both claim to represent the global mean anomaly for the same planet during nearly the same time period. The fact that they are so different is, on its face, proof that something is drastically wrong with the entire system.

The adjustments to the bottom graph only accentuate the differences and results in a plot that has essentially no correlation with the rising CO_2 concentration in the atmosphere. Every time the anomaly decreases and the CO_2 increases the correlation is weakened. The greater the decrease each time, while the CO_2 concentration steadily increases the greater the negative impact on the correlation. This is a global warming theory fatal fail.

If you are not convinced at this point I have a final set of examples that you need to consider. In my opinion this represents the ultimate fatal fail of the GISSTEMP SAT system. In 2019 at <https://data.giss.nasa.gov/gistemp/uncertainty/> a publication from the Goddard Institute of Space Studies (NASA) dated May 23, 2019, "Uncertainty Quantification" was unveiled. The graph, p. 232 and caption, is the key illustrative feature of this publication.

Caption: The production GISTEMP global mean temperature time series with the total (LSAT and SST) 95% confidence interval calculated in our study for annual mean temperature smoothed with LOWESS with 5-year bandwidth. Anomalies are calculated with respect to the 1951-1980 climatology.

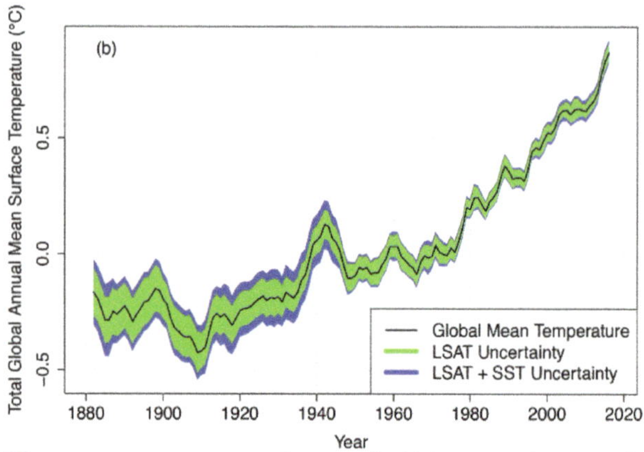

There are a number of things about this "Uncertainty Quantification" that are problematic and/or questionable.

First: The claim in the above caption is that there is a 95% degree of certainty the above confidence interval limits are correct. The confidence limits are illustrated by the green and blue shading on both sides of the black line which represents the annual mean temperature. Quoting the article "The resulting 95% uncertainties are near 0.05^{0}C in the global annual mean for the last 50 years, and increase going back further in time reaching 0.15^{0}C in 1880."

Let's take a look at what the data situation was 140 years ago and then 50 years ago. This will give us a basis for determining how believable the claims of a +/- 0.15^{0}C and +/- 0.05^{0}C confidence interval for those historical periods are.

In the publication by Hansen and Lebedeff (1987) <https://pubs.giss.nasa.gov/docs/1987/1987_Hansen_ha00700d.pdf> describing this temperature estimating system: Figure 1; illustrates the earth, divided spatially into 80 equally sized 'areas'. The 1200 km radius circles represent the areas covered by each temperature monitoring site. In 1870, no portion of 32 of the 80 areas were covered by any temperature monitoring stations. That is 40% of the earths surface. In addition 9 others had less than 1/4 covered by any of the circles, so the actual figure is 45% of the earth's surface had no data representing it.

In 1900 the number of these areas, with no monitoring stations was 20, or 25% of the earth's surface. Additionally in 1870, eight of these regions contained only 1 monitoring station (coverage = <½ of the area) and in 1900 there was only 1 monitoring station in five of these regions.

What the above two paragraphs illustrate is that from 1880 to 1900 the 0.15^{0}C confidence interval was achieved in spite of the fact that 30%+ of the earth's surface had no temperature data representing it (See Fig. #4 at the above link). That means that they needed to extrapolate the existing data across those areas and still maintain an accuracy of 0.15^{0}C. The claim is that their estimate of the earth's surface temperature is accurate to within ⅙ of a degree. How believable is that claim? (Again, if that seems reasonable to you, obtaining the deed to that sea side property on the San Rafael Swell in central Utah should be your highest priority; contact me.)

Authors Note: *According to the above cited document (by Hansen and Lebedeff), the 1200 km circles had a correlation coefficient of .5 in the mid latitudes and .3 in the low latitudes (the low latitudes constitute 40% of the earth's surface). These low correlation coefficients preclude the error limits from being as narrow as claimed. A .3 Correlation Coefficient (r value), yields a .09 Coefficient of Determination (r^2 value), meaning that a maximum of 9% of the relationship being examined can be attributed to the comparisons being made (for 40% of* *earths surface); A .5 r value velds a 0.25 CD for the temperate zones (roughly 51% of the earth's surface).*

The unrepresented areas tend to be far removed from the closest station, with the Arctic, Antarctic, South Central Pacific, Southern Atlantic and Siberia constituting a high percentage of the uncovered area. As the distance from the monitoring site increases, spreading temperatures increases the error.

For the 1951 to 1980 period, with a much more complete spatial coverage of data over the earth's surface, there is an admission that the confidence interval is at least +/-0.6^{0}C, according to the February 12, 2019, publication about the GISSTEMP SAT system we reviewed.

The same qualification applies to the claimed 0.05^{0}C claimed as the confidence interval for Global Annual Mean Surface Temperature from 1968 forward. The admitted error (+/-0.6^{0}C) is 12 times the 0.05^{0}C confidence level and 4 times the 0.15^0 confidence level. Hopefully you understand how troubling this type and magnitude of internal inconsistency is. It represents careless scientific effort, or, in the alternative a strong bias, possibly, even an attempt to deceive.

Authors note: The only way you can explain the above extremely important inconsistency is to recognize that it results from the increased data manipulation involved when the GHCNv4 data set was adopted, and LOWESS smoothing.

For example, if you locate the 1998 to 2018 segment of the above chart (the one with the green and blue confidence interval) you will note that the slope is much steeper and a different shape than the slope for that same period on the chart which is based on the actual numbers. This illustrates the fatal fail of the LOWESS technique coupled with the other data manipulation involved when the GHCNv4 data set was adopted.

Excuse a quick trip into the technical weeds. Computation of the correlation coefficient between the change in the CO_2 concentration of the atmosphere and the change in the anomaly, yields an r value of 0.042 for the 20 years between 1998 and 2018 using the actual numbers from NASA web archive. There appears to be no correlation between the 24.88 ppm rise in the CO_2 concentration and the flat anomaly between 1998 and 2011. In fact the net rise in the anomaly during this period equaled 0.0^0 C.

When the GHCNv3 version of the data was adopted in 2011 the anomaly began a steep rise. The net rise between 2012 and 2018 was equal to 0.31^0C even though 3 of the 7 years actually exhibited a decrease in the earth's surface temperature (again, using the real numbers). When the LOWESS numbers (based on the GHCNv4 dataset) were used, the net rise of the anomaly for the entire 20 year period was equal to 0.53^0C. This amounted to an increase of 0.21^0C in the anomaly over the 20 year period due to a change in the dataset and a 0.32^0C change due to LOWESS smoothing. The combination of the revised data set and LOWESS smoothing results in an extreme distortion of the real picture if actual data is used. (I prefer the original recipe for fudge, preferably not spread thickly over data.)

If you use the LOWESS numbers, the coefficient is equal to .457 using the GHCNv3 data combined with LOWESS smoothing (5yr). This is an example of how changing the numbers changes a result. The correlation coefficient changed by a factor of 10 by the use of the "unadjusted" (but changed) data in combination with LOWESS smoothing.

Rhetorical Questions: Do you think that,"a scientist" should be able to change the data and calculation method in an experiment or procedure whether or not the result is changed? If such changes consistently change the result so it is more favorable to the theory is that suspicious, and possibly fraudulent?

The changed numbers change the results obtained if they are used in other data analysis techniques, because they hide the real values and degree of variation. For example a correlation analysis of these temperature numbers with the CO_2 concentration of the atmosphere will be entirely different (a much better correlation as shown above) than the raw data. The same applies to an analysis of variance (ANOV).

Question: Are the LOWESS smoothed numbers used in determining the claimed narrow confidence interval? According to the caption, the answer is yes. That is a refined, artful and deceptive application of extreme bias.

That last statement comes from my conviction that such a practice is reprehensible. It is not science. If you change the numbers knowing that it will change the results that is "Fraud" (a person or thing intended to deceive others, typically by unjustifiably claiming or being credited with accomplishments or qualities). There is an act with the intent to deceive (these are educated persons; it could not be ignorance; Bias isn't an adequate extenuating defense).(see note at the end of Appendix #1: It may not be fraud if- -.)

Third: This graph uses a 5 year bandwidth for the LOWESS smoothing, meaning that a 5 year rolling average is used, not the actual annual average numerical value of the anomaly. This removes a significant portion of the annual variation you see when the actual data are displayed. It also delays, makes smaller, and even hides changes in the direction of the current trends. This bandwidth is part of the problem addressed above.

Authors note: A suggestion to the authors of this travesty: if you want to totally smooth your results, having abandoned all basic scientific principles, use a 140 year bandwidth so you can have a 0.00^{0}C confidence interval (Your video production studio could make it look really good).

At this point I would recommend that you go to NASA: <https://climate.nasa.gov/vital-signs/ global-temperature/> and refresh your memory of some details of the interactive graph of the annual global temperature anomaly from 1880 to 2021. Note the annual mean display (gray lines and circles) and the "smoothed data" (black line). Rhetorical Question: Do you see the internal inconsistency? The smoothed data line is still going up (in 2017-2021) when the actual anomaly has been flat for 5 years (this false trend line has continued through 2021, though the slope has decreased {the 2021 anomaly is equal to 0.85}). **How soon should we expect GHCNv5?**

The claim is that there is no decrease in the earth's surface temperature. Those claims are false on their face. The further claim is that this is just a temporary decrease. That is an assumption that may or may not be true, unless you fudge another data set that interrupts the downward trend, which is what happened with GHCNv3 data set in 2011 and the GHCNv4 data set in 2019.

Again I ask: How soon should we expect the GHCNv5 dataset ? See <fatalflaws.net> for the answer, under "Evolution of the Anomaly"

Conclusions:

There are two major problems that underlie the Global Warming/Climate Change theory.

The first problem arises because of the basic assumption underlying the theory, which is: All the warming of the earth's surface that has taken place since 1880 is the result of increasing CO_2 levels in the atmosphere. (If you doubt this statement please check out Mr. Al Gore's speech at the January 2023 World Economic Forum. He referred to any deviance from this "principle" as B—S—- at least 4 times).

At the beginning of this book you were exposed to a simple, easily understood comparison between the single dimensional modeling system used to support what I referred to as a ridiculous assumption and a multi-dimensional model or concept. I am sure that some thought I was being harsh and judgmental when I wrote that. I suggested that you continue to read and indicated that I would ask at the end of the work if you still felt that way. Let us review the comparison.

Let us view the contents of a Mason Jar through the prism of a single-dimensional model between Temperature and CO_2 with the basic assumption that, "Any heating will be the result of CO_2 forcing":

We begin with a 1 quart Mason Jar containing 1 pint (473 g) of water at a temperature of 20^0C.

Step A. We add a net of 473 calories of heat as radiant energy from CO_2.

Step B. The result is that the temperature of the water is now 21^0C.

Step C. The conclusion: The CO_2 has caused a 1^0C rise in the water temperature.

Step D. We now add a net of 473 calories of heat as sunlight energy.

Step E. The result is that the temperature of the water is now 22^0C.

Step F. The conclusion: The CO_2 has caused a 2^0C rise in the water temperature.

Does the above make sense? Answer: No. But, remember the controlling assumption: "all heating will be the result of CO_2 forcing of the contents of the Mason Jar," hence the conclusion, no matter how ridiculous; "the heating is all the result of CO_2" forcing of the jar's contents. If the heat came from water, magma, or the greenhouse impact of aerosols the conclusion would be the same, forced by the underlying assumption: "The heating is all the result of CO_2 forcing."

Now let us view the contents of a Mason Jar through the prism of a multi-dimensional model between Temperature and CO_2 with the basic assumption that any heating of the contents of the Mason Jar may be the result of energy from 1. The sun, and/or 2. Water, and/or 3. Magma, and/or 4. Aerosols, and/or 5. Carbon Dioxide:

We begin with a 1 quart Mason Jar containing 1 pint (473 g) of water at a temperature of 20^0C.

Step A. We add a net of 473 calories of heat as radiant energy from CO_2.

Step B. The result is that the temperature of the water is now 21^0C.

Step C. The conclusion: The CO_2 has caused a 1^0C rise in the water temperature.

Step D. We now add a net of 473 calories of heat from sunlight energy.

Step E. The result is that the temperature of the water is now 22^0C.

Step F. The conclusion: Carbon Dioxide has caused a 1^0C rise in the water temperature, and the sun has caused a 1^0C rise in the temperature of the water .

If the heat added in A, or D resulted from sitting the jar in warm water, or on cooling magma, or from the greenhouse effect caused by aerosols, the added heat would be credited to the proper source. That does not happen with a single dimension, myopic modeling system. The Global Warming/Climate Change theory, because of this attribute of single dimensional models, lumps the warming contributions of all other contributing forces together under the carbon dioxide forcing moniker.

When it is broken down it is easily understood, and the "Fatal Fail" forced by the ridiculous underlying assumption is exposed. The following is a quick, simple review of the most salient facts.

The Sun: The spacial relationship between the earth and the sun as well as the tilt of the earth's axis are fixed (unchanging). The temperature of the sun does vary as the number of sunspots and solar flares change during the solar cycles. When there is an increase in the intensity of solar activity over an extended period of time, there is also a change in the temperature of the "quiet sun". Since 1900 both of these factors have increased the amount of solar irradiance that reaches the outer limits of the earth's atmosphere and ultimately the surface of the earth.

As illustrated by the TIM's graph near the end of Chapter #1 there was a gradual increase in the intensity of solar activity beginning around 1900. This increase also gradually increased the temperature of the sun during the "quiet sun" stage of the cycles (shown as the area under the bases of the peaks, which illustrates the increasing solar activity, the solar maximum, and the decreasing solar activity of each cycle).

Solar cycle 19, as illustrated by the above mentioned graph and the graph from the Royal Dutch Observatory shown earlier in the chapter, was the most intense in recorded history. There followed a cycle (#20) of significantly lower activity, though it was still greater than any of the preceding cycles since 1600, then 3 very intense cycles followed by cycle 24 which was lower in intensity with an extended minimum (to at least 2/1/2021).

During this extended period (1900 to present), the increase in the temperature of the quiet sun plus the temperature increase during each cycle increased the solar irradiance reaching and absorbed by the earth's surface. This elevated solar energy has been enough to account directly for 50% of the total increase in earth's surface temperature.

Over the last 75+ years the ozone depletion in the upper atmosphere has added significantly to the amount of x-ray and UV light that has reached the earth's surface (a warming impact). That force is currently being attenuated because of international treaties, reducing then eliminating use of ODS (ozone depleting substances), enacted in the late 1980s. These treaties became fully implemented in 2005. This warming impact has nothing to do with the minor greenhouse gasses. It has increased the suns role in the warming of the earth's surface (a 3rd solar warming mechanism).

According to a publication of the Royal Society of Chemistry article entitled "CFCs responsible for half of Arctic sea-ice loss" by: Jamie Durrani published January 20, 2020: "Now, an investigation led by Columbia University geophysicist Lorenzo Polvani has revealed the extent to which ozone-depleting substances (ODS) contributed to temperature rises and sea-ice loss in the Arctic by direct radiative warming, between 1955 and 2005. On a molecule-by-molecule basis, halogenated organic compounds trap much more heat in the atmosphere than most other known compounds. For example dichlorodifluoromethane (CFC-12) has a global warming potential almost 11,000 times that of carbon dioxide. This means that, although ODS might exist in much smaller atmospheric concentrations than other greenhouse gases, their impact can be just as significant".

The total impact of the above three mechanisms of warming of earth's surface is credited to carbon dioxide because of the single dimension characteristic of the basic model: i.e. The heating of the earth's surface is the result of carbon dioxide forcing.

Earth's Waters: Water is the most widely distributed substance on the earth's surface. The vastness of this surface and the amazing physical properties of water combine to make it second only to the sun as a surface temperature modulating and controlling force. Water also interacts with the relatively small increases in solar radiation which reaches the earth's surface in ways that multiplies the impact of both of these climate control forces.

As water evaporates from earth's surfaces, it carries tremendous amounts of energy (heat) upward in the atmosphere in a form (as chemical potential energy) which cannot be intercepted by the minor greenhouse gases. This results in approximately 75% of all the heat that is lost by the surface of the planet, by-passing a significant portion of the greenhouse potential of CO_2 and other minor greenhouse gases.

Global warming/climate change dogma views water as serving merely to modulate earth's climate by various feedback systems. This myopic view ignores the fact that the evaporation, condensation, precipitation cycle is the force that controls atmospheric pressure and is also the energy source that powers storms. These are climate driving forces, not feedback mechanisms.

During ElNino cycles the amount of water in the atmosphere increases by 1,100 Trillion pounds. That intensifies both low pressure and high pressure systems. The high pressure systems tend to dominate the polar regions during ElNino leading to a significant portion of the temperature increases, and ice melt that has occurred there during the last 38 years.

<u>Heat from Earth's Core:</u> There has been considerable debate about the role of heat from the earth's core and its role in warming the earth's surface. Most of this debate has focused on land based Volcanos. That heat and particulate/aerosol source is inadequate to cause a significant impact (though it undeniably has some effect).

When the land based volcanic heat and aerosol source is combined with the impacts of other earth core heat sources including:

1. Volcanic activity under the seas and polar ice packs.
2. Magma flows onto the ocean floor and intrusions into the junctions of tectonic plates.
3. The heat that moves upward through the earth's crust by convection and,
4. Thermal vents and other core driven heat sources, the true magnitude of this source of heat can be appreciated.

As with the sun, in addition to this heat sources' primary or direct impacts, the indirect impacts which result from the heating of the ocean's surface are highly significant. Increased evaporation and the climate changes which result, have major impacts on the earth's surface temperature.

Particulate and Aerosols: During the last 140 years the amount of particulate and aerosols in the atmosphere has increased greatly. NASA maintains that 90% of the particulate and aerosols are from natural sources and only 10% of these air borne particles are human caused. For the purposes of this work I will stipulate that these figures may be correct, with the added stipulation that the 10% is, 10% of a rapidly increasing total.

Particulate and aerosols play a very important role in attenuating the amount of solar irradiance that reaches the earth's surface. Approximately 15% to 20% of the sunlight that reaches the outer limits of the atmosphere is scattered or reflected back into space by these particles.

The global warming/climate change advocates recognize the role of these particles in attenuating the incoming solar radiant energy. They ignore or greatly underestimate the role of these particles as active agents in the atmospheres greenhouse effect. This means that the impact of these particles on the warming of the earth's surface is mistakenly credited to CO_2 and other minor greenhouse gases.

Carbon Dioxide: This gas is one of the most important substances in our environment because of the major roles it plays. Each year hundreds of billions of tons of CO_2 are removed from the atmosphere by the process of photosynthesis. Most of this CO_2 is recycled back into the atmosphere as organic materials are oxidized but significant amounts are sequestered. In addition billions of tons are removed from the air by 3 non-photosynthetic processes. The net result is that, of the 40+ billion tons of CO_2 released by human activity annually, most of it is sequestered, leaving approximately 11.35 billion tons (2ppm) as the net addition to the atmosphere.

I have a nephew who is a MD with a speciality in Environmental Toxicology. He has, more than once, pointed out to me that: The pH balance that is maintained by the carbon dioxide/ bicarbonate dynamic equilibrium is the buffer system that controls mammalian pH within a range that allows life to exist.

Wildfires and Reforestation: A major contributor to the atmospheric CO_2 concentration increase in the last 30 years has been a significant increase in the size of wildfires. The number of fires has increased slightly because of increased human use of wild lands for many different purposes. In the United States in the last 10 years there have been over 60,000 wildfires per year, with approximately 9,500 of them being started by lightning or other natural events.

In the US, the area burned by the average wildfire increased from 43 acres (1990), to 119 acres for the period of 2009 to 2018. This increase has occurred because of changes in fire control procedures that gradually took place in the late 1980s and became national in 1995 and 1996.

Since 1700 there has been approximately 5.5 billion hectares of forest removed from the earth's surface. Restoring or replacing approximately 250 million acres of properly managed forests, dedicated to the production of building materials could stabilize the CO_2 concentration in the atmosphere by sequestering the excess for hundreds of years. (I devoted two chapters in this work to a thorough, though not exhaustive, treatment of the impacts of Wildfires and the potential of reforestation to control the level of carbon dioxide in the atmosphere. I intend to publish a thorough reforestation plan and cost/benefit estimates for the temperate and tropical zones in 2023)

The second major problem with the global warming/climate change theory is related to how the average mean global temperature is determined.

A vast array of ever changing monitoring systems are relied on for data. The data that are produced are, "out of necessity, edited," changed and processed to make them compatible. There are, in these systems a number of opportunities for various significant forms of bias to be introduced. There is significant evidence that bias has drastically inflated the current estimated increase in the earth's surface temperature. Consider the following:

Quoting from a NASA document you can examine at: <https://data.giss.nasa.gov/gistemp/faq/index.html#q206> Q. Do the raw data ever change, and why do monthly updates impact earlier global mean data?

A. "The raw data always stays the same, except for occasional reported corrections or replacements of preliminary data from one source by reports obtained later from a more trusted source".

The above question and answer is very reassuring. The raw data always stays the same (with an unspecified number of "corrections or replacements"). Still that appears to be reassuring. There is a "But": the answer to the above question suggests that even though the raw data always stay the same, it is not the data that is used.

The raw data are adjusted through a process: The next question and answer reads: Q. "Why do the unadjusted data in GHCNv4 differ from the ones in GHCNv3?"

A. "The 'unadjusted' data shown in the station data viewer are not raw data. They are the 'qcu' data retrieved from NOAA/NCEI". The "unadjusted" data are created from raw data by a series of processes which involve:

1. A collection process from multiple sources. This is part of the homogenization process.

2. An integration process to create daily means. The daily means are the result of the homogenization (integration) of the multiple sources (data sets).

3. The monthly means are generated (the result depending on how missing daily reports are handled).

4. A merge process, of multiple sources for the same location (this involves a human decision and of course will be made impartially?).

So the raw data is adjusted, by a 4 step process into the "unadjusted data". I know that does not make sense, but that is what is described. This question also reveals that the GHCNv3 data was changed into the GHCNv4 data and in the process, a vastly larger collection of data was created. (Elsewhere you will find that several thousand additional observation sites were added in the formulation of the GHCNv3 dataset and nearly the same number of additional data collection sites were added in the formulation of the GHCNv4 dataset).

Finally it is stated that: "Please note that we don't use the 'qcu' data for the GISTEMP analysis but the 'qcf' data that has been additionally processed for homogeneity adjustments" (the 5th step).

The 5th step mentioned above is actually the 1st step (in the time sequence). As data arrives in the system it is compared to previous values from the same site and "neighboring sites". This is done by an algorithm which will adjust or drop the data if it does not fit the pattern relationship (pattern relationship is my term).

Rhetorical Question: How do you add 13,000 additional observation sites into a data base (the GHCNv3 dataset) and maintain the relationship between the limited observation sites used to establish the "normal temperature" (1951 to 1980 average global mean temperature) and a current number of data acquisition points that is many times larger? Answer: You introduce additional significant errors, which you hide by cloaking them in long, wordy, erudite expressions; Then you do it again to formulate the GHCNv4 dataset!

Authors Note: Another thing that is revealed in the above question and answer is that we are now using GHCNv4, or version 4 of a seemingly constantly changing data set. And yes, there was v3 (and v3.1, 3.2). There was also a v1 and v2. Changes made in all predecessor versions are preserved (or increased) in subsequent versions.

When the above is examined closely it becomes apparent that NASA is using a data set and data manipulation method similar to what was used to produce the Berkeley Earth Graph which appears and is analyzed several times in Chapters 9 and 10.

When you contemplate what is contained in the above description of how the data is processed you should understand that, when I said: "the data is dipped in a cooked mixture of cocoa powder, powdered sugar, milk and butter along with other ingredients of your choice, in other words fudged"; I was not, in my opinion, being harsh and judgmental, merely descriptive.

Ask yourself: How do you prevent bias from flooding into such a multi-step, data processing maize? **Someone should study the elements of experimental design so proper data, in proper formats, can be collected by proper equipment, properly placed.**

It is an undeniable fact that, with a limited number of exceptions involving some of the data corrections from individual sites, the corrections made have been preponderantly to correct "cooling errors". This increases the value of the average global mean temperature.

As I wrote previously; in a system involving thousands of sites (currently 28,000+) and hundreds of millions of data points, cooling errors should be very close in numbers and magnitude to warming errors.

They should cancel each other out. In this scheme they do not. As the "cooling errors" are corrected, the process leads to significant increases in the estimated anomaly.

Statistical anomalies (abnormal occurrences) do occur, but not sequentially time, after time, after time, after time, after time, after time. Note: the word time appears 6 times in the previous sentence. Once for each of the versions of the GHCN data sets. Technically a seventh is warranted for the conversion from the NOAA/ NWS data set to the GHCN data set.

The above is why I wrote at the end of Chapter 10: The total data changes and adjustments have amounted to approximately 0.45^0 to 0.56^0C. This means that the true value of the global mean anomaly is somewhere between 0.40^0 and 0.50^0C.

As I have mentioned several times, the global warming/climate change doctrine is founded on a basic assumption that the rising CO_2 concentration in the earth's atmosphere is responsible for the rising global average mean surface temperature. In fact, the doctrine holds that it is totally (100%) responsible for any increase in the earth's surface temperature.

This assumption is the foundation for the global warming theory. It rests on another assumption, which is; there is a strong correlation between the rising CO_2 concentration and the rising mean global surface temperature. To assess the global warming/climate change theory we need to examine this assumption. To make such an assessment, let's begin by examining closely a graph that you have seen several times in the preceding chapters.

The graph below can be found in several places on the web. This version covers the time period from 1880 to 2014 and shows the plots of four of the most reputable agencies that work in this area of science. They are indicated in the body of the graph. An interactive version of this graph exists and was referenced earlier.

First let me draw your attention to several things. They are:

1. In 1880, 3 of the 4 agencies' plots begin at -0.16⁰C (JMA begins in 1890 at -0.27⁰C).

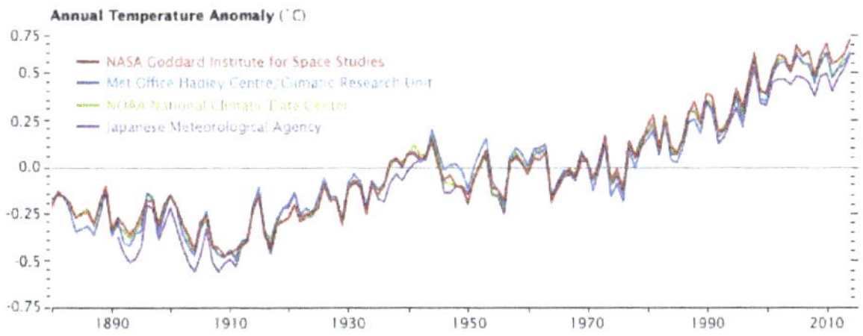

2. The plot lines move up and down several times in the next 30 years to 1910 which is the period that describes a "climatology". The trend is generally downward to the point that by 1910 all the lines range from, -0.45⁰ to -0.51⁰C. This is a -0.3⁰C decrease in the anomaly. During this time period the concentration of CO_2 was increasing and there were 4 short term changes in the anomaly equal to 0.25⁰ to 0.3⁰C.

3. Between 1910 and 1940 (another 30 year period or "climatology") the anomaly climbed back up to the 0.0 line, which is the global average mean surface temperature for the 1951 to 1980 period (this 30 year period serves as the baseline or reference [normal] temperature for the GISS system).

4. During this period (1910 to 1940) the temperature anomaly again varied significantly over short periods of time. These plots show that for the 1st 30 years the correlation between the change in temperature and the change in CO_2

concentration of the atmosphere was negative. During the next 30 years the correlation was positive, but very weak because during the entire 60 year period the anomaly moved up and down frequently while the CO_2 concentration climbed steadily. Each change in the relationship between the anomaly and CO_2 concentration weakened the correlation.

5. During this time period (60 years) the CO_2 concentration of the atmosphere steadily increased from 290 ppm to 311.3 ppm (equals 21.3 ppm, or 17.6% of the total increase between 1880 and 2019).

6. Between 1940 and 1950 the anomaly increased to 0.2^0C, then fell back to -0.2^0C.

The correlation coefficient between the change in the temperature anomaly and the change in the CO_2 concentration over this 70 year time frame was .0593 (extremely weak). For this 70 year time period these figures indicate that there was almost no correlation between the atmospheric CO_2 concentration and the earth's surface temperature. The Coefficient of Determination for this period estimates that less than 1/4 of 1% of the changes in the annual global anomaly could be attributed to changes in the CO_2 concentration of the atmosphere.

During the period of 1951 to 1980 the correlation coefficient between CO_2 concentration and the global mean anomaly was equal to .25. This means that the Coefficient of Determination was equal to .0625. During that period the maximum contribution of CO_2 to the increased global mean surface temperature was approximately 6.25%. This means that other forces accounted for 93.75% of the total warming.

From 1980 to 2018 the correlation coefficient was stronger, largely due to the changes in the “raw data” (not used) when it was converted to the “unadjusted data” which is the basic data set used (through the above mentioned 6 versions, all of which corrected cooling biases — a statistical abnormality). I wrote above: “Statistical anomalies, abnormal occurrences, do happen, but not sequentially time after time, after time, after time, after time, after time.” Note: the word time appears 6 times in the previous sentence. (once each for GHCNv1, GHCNv2, GHCNv3, GHCNv3.1, GHCNv3.2, and GHCNv4)

Again, I realize that in a number of places in this work I have been repetitive or redundant. This is because of the importance of this topic to the future well being of my children and grandchildren (and yours). If we waste our resources, or restrict their use unnecessarily we will greatly limit their ability to enjoy this earth and the bounties and beauty it has to offer. If they live a “poorer” life, the less fortunate will also experience unnecessary limits on their ability to improve their life circumstances.

I would hope you, and everyone would take the time and make an effort to contemplate these things, for your sake, and the sake of humanity. Thank you for reading and considering these things. For more on these topics visit: <fatalflaws.net>

A “Correlation Analysis Primer”

A mathematical analysis technique called “Correlation Analysis” is used to determine the degree of correlation, and estimate the percentage of causation which exists in an analyzed relationship. The degree of correlation is expressed as a value of (r). It may be a positive value if one factor goes up when the other goes up, or it may be a negative value if, when one factor goes up the other one goes down. For example, my weight is positively correlated to how many calories I consume each day. If I have a low level of physical activity, the r value is positive and very close to 1.0 (the + sign is usually not displayed, it is understood).

In 2007 I reached the point that I was a little on the chunky side (at 5'10" and 215 pounds). Since I could no longer claim it was just my heavy bone structure, or muscular build I began a diet regime. From March 1 to December 31, I began a weekly regime in which I added to my weekly diet at least 2 and often 3, twenty-ounce Malts from a local (mom-and-pop) fast food outlet. I also added 1 "double scoop" high quality ice cream (waffle cone) as I made my weekly trip through the orchards I visit.

At the end of the year, after consuming over 100 large malts and over 40 large ice cream cones, in addition to my normal diet, my weight had decreased to 186 pounds (I could again claim, it was my muscular build).

I hear you asking, What does this have to do with correlation analysis between the global mean anomaly and atmospheric CO_2 concentration? Just as the global warming advocates fail to acknowledge the role of the sun, earth's waters, heat from the earth's core and atmospheric particulate in the heating of the earth's surface temperature over the last 110 years, I chose to ignore 1 detail of my strict diet regime.

Beginning March 1 of that year (2007), I began to work toward my goal of hiking every side canyon along a 45 mile stretch of Utah Highway 31 between Fairview and Huntington, Utah (USA). There are 41 canyons that range in length from 2.5 to 11 miles long with elevation gains from the start point to the summit of 1200 to 2800 feet.

Three of those canyons required a 3 to 4 mile walk across rough country to get to a start point. Five of the canyons had some type of a trail into, or along the bottom, the rest were "rough as a cob" to use a rural expression.

If I do a correlation analysis between my calorie intake and weight during a period when I go hiking weekly in rough terrain for 8+ hours, I will find that my weight goes down even though my calorie intake goes up. This will yield a negative correlation coefficient or (-r) value.

If I go hiking only once a week for 4 hours, but increase my calorie intake even slightly, I will still gain weight, just not as much as if I had no physical activity. In that instance the r value would be less than 1 but greater that 0 (it may be 0.5). If, in this circumstance the 0.5 correlation value is squared (0.5 x 0.5) I get a value of r^2 = 0.25. This number, .25 is called the "correlation of determination" which is an estimate of how much of my weight gain is because of my increased calorie intake. In this instance, the .25 r^2 value tells me that my weight gain is limited by some factor other than my calorie intake.

Just in case there are any doubts about how my "strict diet regime" is related to the basic point of this work let me succinctly state it. Correlation does not prove causation.

In the rather detailed description of my "strict diet regime" above I related how my calorie intake was increased over the 10 month time period. I assumed that my weight loss was caused by the increased calorie intake (my basic assumption). In actuality, the weight loss was caused by the increased calorie usage, attributable too frequent, strenuous hikes which "burned a lot of calories". In short, my initial inferred assumption had no valid relationship to the true cause of my weight loss.

My frequent, strenuous hikes did have a strong correlation: A known mechanism to cause my weight loss: And would be repeatable. Hence it would be scientifically appropriate to attribute my weight loss to the increased exercise. If I considered my initial inferred assumption, that any weight loss would be attributable to the calories I consumed to be "settled science" I would end up attributing my weight loss to my higher calorie intake.

In this work I have made the case that the inferred correlation between a rising CO_2 concentration in the atmosphere and an increasing earth's surface temperature only appears to be strong if you did not run the proper mathematical analysis. If you did the proper mathematical analysis (using non-fudged numbers) and examined closely the other requirements necessary to determine if the inferred correlation was actually the cause, the invalid nature of the basic assumptions are obvious. When the impacts of the 4 major drivers of earth's surface temperature are properly credited as causative forces the picture is vastly different and clear.

Let me say that again. When the impacts of the 4 major drivers of earth's surface temperature are properly credited as causative forces the picture is vastly different. Failure to treat "nature" as the beautiful, dynamic, system it is, results in the ultimate fatal fail of the Global Warming/ Climate Change theory.

In the third paragraph above I wrote: My frequent, strenuous hikes did have a strong correlation, a known mechanism to cause my weight loss, and would be repeatable. This sentence mentions the 3 requirements before correlation can be considered to be a cause of a result (Jacobs rules).

First; there must be a strong correlation. In many fields of science the correlation coefficient must be .9 or higher before one can be confident at a 90% or 95% level. For example, in plant nutrition work if the change in the plants growth and the application rate of a particular nutrient element or form, yields a correlation coefficient below .9 (r^2 value of .81) you can't confidently say that the change in growth is caused by the application. This is because there are a large number of variables which may influence the result, most of which were not "controlled".

Second: There must be a known, definable and testable mechanism of how the force in question can bring about the result. In the case of CO_2 and other minor greenhouse gases there is a definable mechanism by which they may contribute to the greenhouse effect of the atmosphere. The question is; How much of a net contribution can 400 ppm make? (See Appendix #2 p. 267) This paragraph leads to 2 additional question sets. When the claim is, that 100% of the warming of earths surface results from CO_2 forcing of earth's surface temperature, a number of additional questions flow from that claim:

a. Could the claim made be accurate? (My answer) The total increase in the CO_2 concentration in the atmosphere is equal to 121 ppm. That is 39% of the current level of 411 ppm. The total concentration of CO_2 in the atmosphere now (10/10/2021) is 411 ppm (down from its peak of 419 ppm in June). At 411 ppm the concentration of CO_2 is equal to 10.1% of the water vapor present in the atmosphere on a weight basis. On the basis of the number of molecules per cubic meter of atmosphere, the water vapor is 24.4 times as high as the CO_2 concentration. In addition, the liquid and solid states of water are known to be very active in driving the greenhouse effect and they are approximately 10 times the water vapor concentration.

The vastly greater concentration of water in the atmosphere means that a very slight increase in the total water vapor content of the atmosphere (approximately 12.1 ppm) will increase the greenhouse effect more than the 121 ppm increase in the CO_2 could possibly cause. During ElNino periods the quantity of water in the atmosphere increases by approximately 1,100 Trillion pounds. This is why the Lawrence Livermore Laboratories article cited in this work has indicated that the ElNino cycle over the last 38 years has caused at least 40-50% of the ice melt in the Arctic.

When the fact that, 75% of the energy that leaves the Earth's surface, does so as potential chemical energy (the "latent heat of vaporization" not subject to interception by CO_2) is considered the role of CO_2 in the greenhouse effect is further diminished.

b. What are the conditions under which it could be accurate? (My answer) The claim above could only be made accurate if the concentration increased by many fold what it is currently. As CO_2 increases in the atmosphere it limits incoming solar radiation which would buffer any heating impact. At least two of waters absorption bands lie within the visible and near infrared spectrum where the energy per photon (packet of energy) is much higher than in the infrared bands where CO_2 absorbs. In addition any heating of the surface increases evaporation of water which transports more heat away from the earth's surface in a form that CO_2 can't intercept. The bottom line is that as long as there is water on the earth's surface and in its atmosphere, CO_2 (and other minor greenhouse gasses) could never control the greenhouse effect. They can and do make a small contribution (<10%).

c. And do those conditions exist in the defined environment? (My answer) Obviously the conditions of earth's surface and atmosphere prevent the current, or foreseeable concentrations of CO_2 from ever being responsible of 100% of the earth's surfaces warming.

The mechanism is that these gases do absorb radiant energy that is given off by the earth's surface. These gases absorb only in limited wavelengths of the radiant spectra and their absorption is concentration dependent. That limits their contribution to an extent that their influences on the earth's surface temperature is minimal. The following is a brief summary of these limitations.

A. Carbon dioxide is a symmetrical molecule composed of 1 atom of carbon and 2 atoms of oxygen. It contains 2 carbon:oxygen double bonds and at atmospheric temperature, within the range that will allow life to exist, it does not associate with other CO_2 molecules or other gases. These characteristics result in a gas with very limited, net, greenhouse potential. (See Appendix 2 p. 267)

For more on these topics visit: <www.fatalflaws.net> or email the author at: <earljseeley@gmail.com>. Submit comments, questions, or insights by email. The rules are simple: 1. Keep it simple, but not too simple, 2. Avoid the use of profanity, 3. Treat all with respect 4. Make no threats to anyone (zero tolerance policy is in effect), and 5. Realize that at the age of 79 my physical capacity is lessened, and I still have not hiked all the trails I would like, so my response will be a little slow at times and questions and answers will be returned as time permits.

My heart felt thanks to you for being interested in this issue of seminole importance and giving me an opportunity to communicate my views and some of the underlying science.

Photo: While my Wife and I were spending some of the most rewarding times of our lives with the Navajo, Hopi, and Apache peoples we were invited to a cookout by a family who lives on the Northwest side of Gray Mountain a few miles southwest of Cameron Arizona.The evening was wonderful with fresh steaks, vegetables grown in a home garden and the works. After supper we took a late drive to the top of the mountain and observed the sunset. The mountains you see in the distance are the upper reaches of the North Rim and Echo Cliffs just below the dam at Page Arizona. What struck me about this view was the almost perfect spectrum of dark to bright white light. On our way off the mountain top to the truck I was struck by the importance of light. There was none and we all wished for more. Contemplate that, as you think of this work. EJS

Appendix #1: "The Normalization of Deviance and Groupthink"

In addition to the natural complexity of the control and modulation system that determines the surface temperature of the earth, this is the best answer I have been able to find and think of for the question: Why is the Global Warming Theory so widely accepted, and the purveyors of this deception so widely respected? The global warming theory is a simple answer. It is much too simple!

Following the explosion and breaking apart of the Space Shuttle Challenger a series of investigations into the cause of this disaster were conducted. In the course of these investigations a NASA report entitled "The Cost of Silence" was published.

The report pointed out how the normalization of deviance played a role in this disaster. Vaughan's Normalization of Deviance states:

"Social normalization of deviance, means that people within the organization become so much accustomed to a deviation that they don't consider it as deviant, despite the fact that they far exceed their own rules for the elementary safety" Diane Vaughan, 1996 (from NASA report: "The Cost of Silence": Normalization of Deviance and Group think; Senior Management ViTS Meeting)

The same principles are applicable to the adoption of the global warming doctrine, and the normalization of practices that depart significantly from earlier scientific principles in order to "prove the theory" and then declare it "settled science".

For example, as discussed in this work; It is a fact that the increase in carbon dioxide concentration and a slight increase in the measured mean temperature of the earth have been weakly correlated.

Over time, through widespread promotion and intolerance of dissent, the uncontested declaration that parallel variation "proved the theory" became widely accepted. In fact, that was not, and is not the case, because parallel variation is not enough to prove cause and effect.

Yet the declaration of the "proof" of the theory was followed by a period of debate over how strong the "proof" was. Then, because the loudest voices were on the global warming side of the argument it became "settled science".

Beneficiaries were; Academia [with notable exceptions who were silenced]; The media {it makes a good story for the 24 hour news cycle and Saturday morning cartoons}; politicians who positioned themselves to take advantage of an emotion-based movement, and many who figured it was a tremendous opportunity to accumulate wealth.

The above quote on normalization of deviance comes from a very sad circumstance; The discussion about what mistakes were made and how those mistakes were allowed, in the lead up to the explosion of the space shuttle shortly after takeoff, with the loss of the entire crew's lives.

The key is the phrase: "people within the organization become so much accustomed to a deviation that they don't consider it as deviant". We all have human failings. Some of the greatest are rationalization, myopic views of what is, and an ego limited ability to recognize when we have been, or are wrong.

The following from: <http://www.geocities.ws/oralcompgroupthink/symptoms.htm> a discussion of "Normalization of Deviance and symptoms of GroupThink," is instructional. Slight adaptation of the 8 symptoms of group think in the rocket launch environment to fit the global warming/climate change situation should be easy enough for anyone capable of critical thought.

Authors Note: The following is, in my opinion, an explanation of how and why the global warming theory has developed into a massive secular religion, that is rapidly making the transition into the position of "The State Sponsored Religion".

It is now evident that the intolerance of dissent is increasingly becoming not just an academic and social force, but that it is being promoted within the educational system as well as in governmental directives, policies, regulation and by Social Media "Platforms".

The current political environment is a manifestation, in many ways, of the 8 symptoms of GroupThink. For years we have been observing the normalization of deviance in the name of diversity, which is now being turned on its head into intolerance of anything but one mindset and thought pattern.

Janis lists eight specific symptoms of a group undergoing GroupThink: *The Marker Felt Font is Janis's description of the symptom*; The, As to Symptom #; is my comment.

1. **Illusion of Invulnerability: Experts in a group will feel that nothing can go wrong after making a decision, because they "are the best" and therefore the decision they made cannot go wrong.** As to Symptom #1: Bias is to be avoided at all costs in safety and scientific research. Arrogance leads to the worst kind of bias, because; "We are the best" gives license to bend or ignore the rules.
2. **Belief in Inherent Morality of the Group: "Experts in a group will believe that the decisions they are making will be morally**

correct". As to Symptom #2: In the case of global warming/climate change experts, not only do they presume to be morally correct, they are, in their own collective minds, morally superior to the point that they have not only the right, but a duty, to shut down and punish dissent.

3. **Collective Rationalization: Members discount warnings that their thinking may be irrational.** As to Symptom #3: Once the step is taken to attribute all of the warming of the earth to three or four minor greenhouse gases and you have signed on with all the force of your faith and reputation (or utilized the movement to gain tenure or wealth), you can't admit that your faith and newly strengthened reputation is built on faulty assumptions or incomplete data and/or facts.

4. **Out-group Stereotypes: The members will all have the same stereotype of the problem.** As to Symptom #4: The current global warming/climate change groupthink has been built over 50+ years of environmental, institutionalized thought, and yes, even mind control patterns. When you start with theme-cartoons for toddlers, reinforce basic thoughts and patterns of thinking throughout K-12, and then continue with outright indoctrination in college and graduate school you are guaranteed that a very high percentage of your subjects will have the same thinking patterns. Add the basic components of intolerance for dissent, declare it settled science and you have accomplished your goal. Very

few are capable of anything else. At any point, the question of; "What are the weaknesses or problems with this theory can't be asked or even thought about, because it is "settled science," demanding **blind faith** of everyone (the Prophets[1] can't be questioned).

5. **Self-Censorship: "When a specific expert is asked what to do about a problem, the expert will not give an answer directly contradicting the group's wishes".** As to Symptom #5: After all, it is "settled science" and you have witnessed what happens to dissenters, they are defamed, persecuted and destroyed.

6. **Illusion of Unanimity: Even if some individuals of the group privately do not consent to the decision made, the individual will not speak out. The individual's silence will then be perceived as consent.** As to Symptom #6: No explanation required, right! We have witnessed the cost of dissent.

7. **Direct Pressure on Dissenters: If an individual does speak out against the will of the majority of the group, all other members will attempt to quell the "dissident".** As to Symptom #7: I remember Dr. Gray of the National Hurricane Center and a number of others; You may not remember them, they were ostracized, defamed, denied grants and promotion, and/or terminated; in short, they were destroyed because they knew and defended the truth.

8. **Self-Appointed Mindguards/**Prophets[1]: **The leader of the group of experts will refuse to hear any argument contradicting his own wishes or the will of the majority.** As to Symptom #8: Self-Appointed Mindguards: We have witnessed this process grow and expand to where the numbers of "leaders" and their reactions to any argument contradicting their own is vast in extent and extreme in the intensity of intolerance. Add to the eight symptoms of GroupThink the economic and power benefits that flow to the "Leadership Class" and you have a self-perpetuating, action, mind, and thought control system that the Stalinists and Maoists of the 1940s, 50's, 60's and 70's could only have dreamed about. Then add Google's social credit scoring abilities (referring to what they have developed for China) and you have George Orwell's 1984 on steroids; by 2025, or 2030?

[1] You will note that I have used certain specific religious designations applied to the leaders and followers of the global warming/ climate change religion. I mean no disrespect toward any religious authority who sincerely believes and teaches the doctrine of a true religion (i.e., one who believes in and worships a higher power than man). I do feel that, self appointed leaders of a secular religion, designed for nefarious purposes, do not deserve such respect.

Appendix: *It may not be fraud

**It may not be fraud, if: I am not an attorney or constitutional lawyer but: A criminal act such as "Fraud" requires two things: First there must be an action and: Second there must be intent, which requires that the person committing the act knows what they are saying or doing is wrong. It could be argued that the persons doing and saying the things marked with (*) lack an understanding of the knowledge that is available in these circumstances, hence, though the actions are wrong the lack of knowledge provides some level of defense. In other words, a little ignorance may give a little defense.*

Appendix #2

The Ultimate "Fatal Fail" of the Global Warming/Climate Change Theory

The title of this appendix arrises because of some incongruence identified in the chart below from the US EPA identified as to source and date downloaded. Immediately below the chart, you will note that the time period is 1750 to 4/2021.

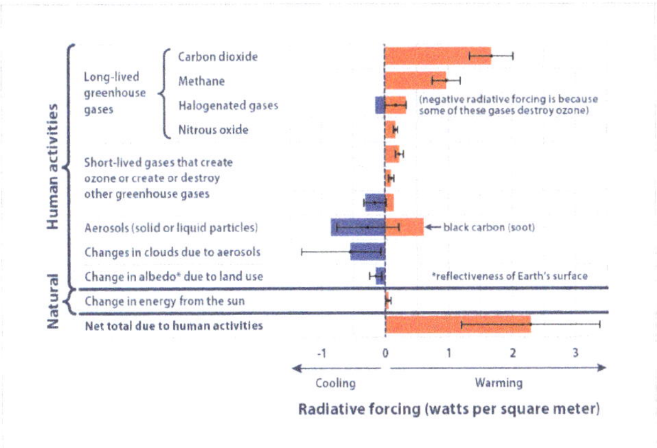

Figure 2. Radiative Forcing Caused by Human Activities Since 1750

Web update: April 2021 - Downloaded 11/4/2021 from US epa.gov

The first and most important of the incongruent facts indicated, are in the bars representing carbon dioxide, methane and nitrous oxide. For example, the bar representing carbon dioxide indicates that the radiative forcing attributable to CO_2 is equal to approximately 1.7 watts per square meter. The value for Methane is equal to approximately 0.9 watts, and Nitrous Oxide is equal to approximately 0.3 watts per square meter.

This group of 3 gasses, purportedly, are responsible for an increase in the greenhouse effect equal to 2.9 watts per square meter (W/m^2) of earth's surface (That is 3.42 times the amount of excess energy absorbed by the earth's surface according to the UNIPCC and NASA's figures. {another inconsistent figure or incongruence}) .

More important; there are no blue bars, or cooling impact shown to the left of the zero line. The incongruence in these representations originates from the fact that, <u>there is an invalid implicit assumption involved</u>.

The implicit assumption and resulting incongruence mentioned above according to this chart is: The increase in the atmospheric carbon dioxide (CO_2), Methane (CH_4), and Nitrous Oxide (N_2O) that has resulted from human activities since 1750, has not resulted in any increased absorption of incoming infrared radiation by these atmospheric components. <u>**If it had, less of the infrared solar radiation would have reached earth's surface, resulting in an offset (represented by a blue bar), equal to, or greater than, the alleged greenhouse impact of the increase in the concentration of these 3 gasses.**</u>

The last sentence of the preceding paragraph is a bold, but absolutely correct statement. The incoming intensity of infrared radiation in the absorption bands of these three gasses is vastly greater than the intensity and energy contained in the absorbable bands given off by earth's surfaces.

That means that as the solar irradiance moves toward the earth it is of greater magnitude so it will excite a higher percentage of the CO_2 molecules. This will decrease the intensity of these bands of radiation as the irradiance moves toward the surface. The irradiance intercepted will be released in all directions, some toward the earth, but mostly back into space. The portion that is released toward space will be greater at the higher altitudes and that released toward the earth will be greater closer to the surface. This is a cooling effect, which is the reverse of the greenhouse effect.

The opposite of what is described in the paragraph above will occur as irradiance is released from earth's surface. This is a warming impact (the greenhouse effect), which is the reverse of what is described above.

Because the temperature of the earth is approximately 57^0F. (on average), the total infrared irradiance released from earth's surface will be significantly lower than the solar irradiance that moves toward the earth's surface. This means that the net effect of the two processes described in the paragraphs above will be a cooling effect. As the CO_2 concentration of the atmosphere increases this net cooling effect will also increase.

There are three relevant ranges or bands of solar radiation reaching earth's surface that are important to this discussion. The first is the ultraviolet band which constitutes slightly more than 8% of the total solar radiant energy. The visible spectrum makes up 42.3% and the infrared makes up 49.4% of the energy spectrum.

The intensity of the solar radiation at the outer limits of earth's atmosphere is approximately 1,360.5 W/m^2. This means that there is 108.8 W/m^2 of ultraviolet radiation, 575.49 Wm^2 of energy in the visible portion of the solar spectrum and 672.1 W/m^2 of energy in the infrared portion of the spectrum at the outer limits of the atmosphere.

Because of the position of the sun and the rotation of earth, there is only ½ of the globe that is receiving direct sunlight at a time. There is also an adjustment required because of the tilt of the earth which causes the seasons. Also the reflection, absorption, and re-radiation of sunlight by clouds and water vapor further reduces the amount of sunlight that falls on the "average square meter" of the earth's surface.

Both NASA and the UN's IPCC maintain that 24% of the solar radiation that arrives at the outer limits of the earth's atmosphere is eventually absorbed by the earth's surface. That means, on average, the infrared energy reaching earths surface would be approximately (672.1 W/m^2 x 0.24 =) 161.28 W/m^2.

The reader probably recognizes that there is a disparity between the figures in the above 2 paragraphs (24% vs. 25%). These are "best estimates," so some difference should be expected. I have used the 24% figure in this Appendix (an arbitrary decision which does not significantly impact the results of this analysis).

The Bottom line is: The increase in the atmospheric CO_2 that has occurred since 1750 is equal to the difference between 280 ppm and 411 ppm, or 131 ppm. (The earlier start date of 1750 results in a 10 ppm greater increase in the CO_2

concentration than the use of the 1880 date, normally considered the beginning of the industrial revolution. Between 1700 and 1920 approximately 4 billion hectares (9.6 billion acres) of forest were lost world wide. This is the probable cause, of 10 ppm rise in the atmospheric CO_2 between 1750 and 1880. {See Chapter 9}).

Again the implicit assumption is that the increased CO_2 in the atmosphere has not absorbed any of the infrared radiant energy from the sunlight moving toward earth. If it did there would be a cooling effect that would reduce, or totally off-set the 2.9 watts per square meter attributed to these 3 gasses. That should be represented by blue bars to the left of the 0 line in the above subject chart. The fact that there are no blue bars is a manifestation of unidimensional thinking or extreme bias.

The amount of the impact of the 131 ppm of increased CO_2 on the percentage of infrared radiant energy that arrives at the outer limits of earths atmosphere vs what reaches the earth would depend on a number of variables. Those variables include:

First: The magnitude of the increase of CO_2 in the atmosphere would be the most important variable. We know that this increase has been approximately 131 ppm since 1750.

Second: The length of the path through the atmosphere that the sunlight must travel would be the second most important variable. There is no indication that the distance from the atmosphere's outer limit to the earth's surface has changed since 1880 (or 1750), so that variable has remained the same.

The Chart below (p. 272) shows the absorption spectra of the natural greenhouse gasses in earth's atmosphere (does not include liquid or solid state water). Note that at approximately 2.8

ABSORPTION SPECTRA FOR MAJOR NATURAL GREENHOUSE GASES IN THE EARTH'S ATMOSPHERE

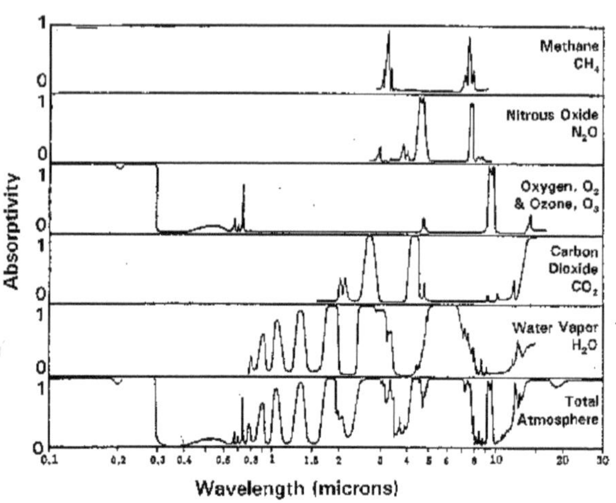

[After J. N. Howard, 1959: *Proc. I.R.E. 47,* 1459; and R. M. Goody and G. D. Robinson, 1951: *Quart. J. Roy. Meteorol. Soc. 77,* 153]

microns (2800nm), there is an absorption band of CO_2. There is also one centered at approximately 4.3 microns (4300nm). (Hopefully this chart will help you understand the significance of the third and fourth variables which are important to these discussions.)

Third: Are any of the bands that CO_2 absorbs, in the same area of the spectrum as water vapor? The answer to that question is, yes, one of the two major absorption bands of CO_2 is centered at 2.8 microns, (2800nm). Water has a very strong absorption band centered in this area of the radiant energy spectrum. Water vapor and CO_2 will remove much of the suns radiant energy in this band before it gets to the earth's surface. A significant amount of this energy will be absorbed, then re-radiated toward space. The 32% increase in the atmospheric CO_2 since 1880, adds to this cooling impact.

Because of the much greater energy content of each photon of radiant energy in this band (2.8 microns) vs the energy of irradiance in the 4.25 micron band, the cooling impact of the 32% increase in CO_2 here is highly significant in determining the net impact of increased atmospheric CO_2.

Fourth: The second major band of radiant energy absorption of CO_2 is centered at 4.25 microns (4250nm). This radiant band is long wavelength radiation with a significantly reduced energy per photon of radiation. The radiant energy that CO_2 absorbs in this portion of the spectrum will be, in terms of its energy content, less than that of the other band discussed above, but it will increase the cooling impact of this mechanism. It will change the net balance we have been discussing toward a greater cooling effect.

The absence (in the graph displayed on p. 267) of any indication of a negative component to carbon dioxide, methane or nitrous oxide forcing indicates that there is something basically wrong with the assumptions, displays, and methods used to arrive at this information. This is a bold statement, but, it is an accurate assertion.

The Bottom line is: The increase in the atmospheric CO_2 that has occurred since 1750 is equal to the difference between 280 ppm and 411 ppm, or 131 ppm.

Again, the implicit assumption is that the increased CO_2 in the atmosphere has not absorbed any of the infrared radiant energy from the sunlight moving toward earth. If it did there would be a cooling effect that would reduce, or totally off-set the 2.9 watts per square meter attributed to these 3 gasses. That should be represented by blue bars to the left of the 0 line in the above subject chart.

The fact that there are no blue bars is a manifestation of unidimensional thinking or extreme bias. (Note: It is apparent that the authors of this document have totally disregarded the fact that the latent heat of vaporization moves most of the heat that leaves the earth's surface as potential chemical energy from the earth's surface to the mid levels of the atmosphere. This heat can't be intercepted by the minor greenhouse gasses until it is released by the condensation process that occurs well above the earth's surface. From that point this energy is much more likely to be radiated into space. This energy is commonly classed as energy from the atmosphere when it is in reality energy from the earth's surface.

I have included this Appendix as an example of the inadequacy of uni-dimensional models and thinking patterns when natural processes are examined. I could literally offer you hundreds of examples of this deficiency and how it results in Fatal Fails of the Global Warming Theory.

I will offer you one additional example as Appendix #3.

Appendix #3: The Problem of Uni-Dimensional Thinking and Models

At: <https://skepticalscience.com/water-vapor-greenhouse-gas.htm> you will find an article that purports to expose a "Climate Myth" written by Tim Ball. I am using this article because it is typical of a vast volume of the literature that defends various aspects of the Global Warming Theory.

First: the quote from Mr. Ball: "Water vapor is the most powerful greenhouse gas":

"Water vapour is the most important greenhouse gas. This is part of the difficulty with the public and the media in understanding that 95% of greenhouse gases are water vapour. The public understand it, in that if you get a fall evening or spring evening and the sky is clear the heat will escape and the temperature will drop and you get frost. If there is a cloud cover, the heat is trapped by water vapour as a greenhouse gas and the temperature stays quite warm. If you go to In Salah in southern Algeria, they recorded at one point a daytime or noon high of 52 degrees Celsius – by midnight that night it was -3.6 degree Celsius. [...] That was caused because there is no, or very little, water vapour in the atmosphere and it is a demonstration of water vapour as the most important greenhouse gas." (Tim Ball) Note: the spelling of vapor is different because I believe Mr. Ball is Canadian or British.

In this article Mr. Ball is pointing out how very low levels of atmospheric water vapor allows the temperature to fall very rapidly when the water vapor's role is lessened by low humidity conditions. He uses an example of what happens in Salah, Southern Algeria where the relative humidity is often very low. (Note: the concentration of CO_2 in the atmosphere is quite uniform across the entire globe and at the time was somewhere between 400 and 419 ppm).

Mr. Ball's point was that even though the CO_2 concentration was over 400 ppm (3.3 times the level of the 121 ppm increase during the industrial period), the temperature dropped from a noon time high of 52^0C (125.6^0F) to -3.6^0C (25.52^0F) by midnight. That was equal to a 100.08^0F temperature decrease in 12 hours (Most of this temperature drop would have occurred near sundown to midnight; in six hours or less).

The following quote (from the above referenced article) illustrates what I have written about the follies of uni-dimensional thinking and models: "Studies show that water vapor feedback roughly doubles the amount of warming caused by CO_2. So if there is a 1^0C change caused by CO_2, the water vapor will cause the temperature to go up another 1°C. When other feedback loops are included, the total warming from a potential 1°C change caused by CO_2 is, in reality, as much as 3°C".

If you take the above at face value, the 1.1^0C current maximum temperature anomaly becomes 0.37^0C ($1.1^0C/3 = 0.37^0C$). But instead of multiplying the forced temperature increase, the evaporative cooling of the earth's surface described and discussed in this work (Chapter 5) moves 75% of the heat that leaves the earth's surface into the mid-levels of the atmosphere in a form that by-passes much of the potential impact of CO_2. This diminishes, instead of multiplying, the forcing effect.

Author's note: The referenced studies involve atmospheric heat flux models based on satellite information which lumps the heat of condensation of water into the category of "radiation from the atmosphere". This hides the fact that, most of that heat left the earth's surface as potential chemical energy resulting from the latent heat of vaporization of water. When you use multiplication because of a myopic view of nature, when division is appropriate, that is called a mistake! This article is typical of the so-called denier debunking articles that are frequently seen.

Additional note: Again we see the impact of uni-dimensional thinking in the fact that Mr. Cook ignores the role of evaporative cooling of the earth's surface and near earth atmosphere and what is discussed in Appendix #2. The net impact of the minor greenhouse gasses is what is important.

The first three sentences of the "debunking answer" reads: "Water vapour is the most dominant greenhouse gas. Water vapour is also the dominant positive feedback in our climate system and amplifies any warming caused by changes in atmospheric CO_2. This positive feedback is why climate is so sensitive to CO_2 warming." So, there is agreement that water vapor is the dominant or most important greenhouse gas.

The rebuttal by Mr. Cook is an admission that climate change advocates recognize the role and status of water vapor in the greenhouse effect. There are a number of myopic problems in this rebuttal. A partial list would be:

First Mr. Cook writes: "If extra water is added to the atmosphere, it condenses and falls as rain or snow within a week or two. Similarly, if somehow moisture was sucked out of the atmosphere, evaporation would restore water vapor levels to 'normal levels' in short time." The last portion of this sentence is partially correct in that; as water leaves the atmosphere it is replaced by evaporation. It is not "sucked" out of the liquid state, it evaporates, taking with it the "heat of vaporization," in response to the negative water potential gradient.

The use of the phrase 'normal levels' in the above paragraph, indicates that the complexity of the system is ignored or not understood. (The amount of water in the atmosphere is not a fixed quantity. It can and does vary widely by thousands of trillions of pounds in response to changing surface temperatures as in ElNino and the North Atlantic Oscillation periods, and other circumstances as discussed in Chapter 1).

That heat is the energy that then rises upward in the atmosphere as potential chemical energy, to the altitude at which the temperature is low enough for the water to condense. This is the heat that I refer to above and which global warming advocates ignore <u>(it represents 75% of all the heat that leaves the earth's surface and it</u>

avoids much of the greenhouse potential of the minor greenhouse gasses). This is why I wrote above: Author's note; When you use multiplication because of a myopic view of nature, when division is appropriate, that is called a mistake.

The first portion of Mr. Cook's sentence above which reads: "If extra water is added to the atmosphere, it condenses and falls as rain or snow within a week or two," is another example of the myopic, single dimension thinking patterns of the global warming advocates. The total volume of water (in all 3 states: liquid, solid and gas) in the atmosphere varies greatly over time, it is not a fixed quantity.

Global Warming advocates, including NASA stress certain "feedback" systems and ignore others. They also ignore the roles of water, where it acts as a primary driver of climatic patterns, particularly where these climate changes are not the result of human activities.

Some of these climatic patterns, such as the ElNino cycle, and the North Atlantic Oscillation result in substantial increases of the amount of water in the atmosphere over long periods of time demonstrating that the above sentence fragment is untrue. It should be noted that water removes from the atmosphere a significant portion of the CO_2 that is released by human activities. That removal directly impacts (decreases) the role of human caused releases of CO_2. (Another very important impact ignored by global warming advocates.)

As an example of what I mean by the last sentence in the 2nd paragraph above, consider the November 5, 2018 publication by staff and collaborators at the Lawrence Livermore National Laboratories which demonstrated that 40% to

50% of the ice loss in the Arctic during the last 38 years can be traced to the ElNino/LaNino cycles.

This is not a human caused ice pack loss. It is not connected to the atmospheric CO_2 concentration. Yet, it has been and still is attributed to atmospheric CO_2 forcing.

Arctic ice loss has resulted from several forces that have caused direct radiative warming of the Arctic. These causative forces include:

1. A decreased albedo (reflectivity) of Arctic surfaces because open water does not reflect radiant energy like snow and ice surfaces.
2. The prominent high pressure areas around the Arctic Circle, caused by the prominent and persistent, low pressure areas in the tropics during ElNino, add to and multiply the Arctic warming. The high pressure areas, result in a higher percentage of the solar radiation reaching and warming the earth's surface. This impact results from reduced cloud cover and lower humidity, especially during the long day-length periods of the Arctic late spring, summer and early fall.
3. The decrease in the ozone levels in the stratosphere has also caused increased radiative warming as discussed earlier in this work. Dr. Polvane in cited peer reviewed work, specifically stated that the ozone depletion caused ice loss in the Arctic, "was a result of direct radiative warming". As reviewed earlier in this work between 40 to 50% of the ice loss in the Arctic over the last 60 years has resulted from ozone depletion driven increases in ultraviolet solar radiation reaching the earth's surface. This is a human caused effect, but is not related to CO_2 forcing of the climate (and has been

successfully reversed by international treaty "The Montreal Protocol").

The single dimension climate models in use, erroneously lump these warming influences, under the carbon dioxide forcing moniker. (Please take note: Mr. Cook and other global warming advocates.)

An additional example of the same problem of the Global Warming doctrine is the increased solar irradiance reaching earth's surface over the last 78 years (a natural variation force). This increased irradiance has contributed a great deal to the warming of earth's surface and ice loss in the Arctic and Antarctic regions.

These examples of how the uni-dimensional model approach, which lumps all of the warming that occurs to the earth's surface, under the carbon dioxide forcing moniker are fatal fails of the theory.

Is it time to ask: Is global warming/climate change a Fraud? How would you answer that question?

Authors Pertinent Bio-Information:

Earl J. Seeley, BA, MS, PhD (Age 79): He was born and grew up in an economically depressed area of Central Utah (USA). He worked his way through college, with the goal of becoming a basic scientific researcher. His BA degree was received from Brigham Young

University with a major in Horticulture and a minor in Botany, with substantial chemistry, and soil science course-work. That was followed by a MS degree in Horticultural Mineral Nutrition and Bio-chemistry from Utah State University.

Following his MS work he was employed as a USU extension agent, and an environmental consultant for a company building power generating facilities in central Utah for four years. He then returned to USU, finished his PhD and joined the Research and Graduate Faculty at Washington State University in 1975.

He was promoted to Associate Professor in 1980. In June of that year he left full-time employment at WSU and joined a private partnership, to develop and manage a large orchard in north-central Washington State.

He continued, as an Adjunct Professor at WSU from 1980 to 1986, then as Adjunct Professor at Brigham Young University from 1987 to 1999. During the last 38+ years he has been a general manager for several large agricultural enterprises and an agricultural management consultant.

Basic research, in both laboratory and field, has been an area of emphasis in his life for over 57 years. Research into the impacts of air pollutants, water stress, heat and cold stress on photosynthesis and cropping patterns, and control or remediation of those effects were central to his work. Additional significant efforts included, irrigation water use efficiency, integrated pest management, and pest control systems and safety.

Frost control research, through the use of wind machines, or combinations of wind machines and under-tree micro-irrigation systems, was begun in 1978. His integration of various technologies in these areas has been widely adopted in the tree fruit, nut, and vineyard industries. Central to this effort, has been the need for, and understanding of, the multidimensional nature of the environmental variables involved in the control of earth's surface temperature, on local, regional and global scales.

The above enumerated experiences, serve as a unique foundation for the author to address and write on the subjects you will find in this work. It is meant to be read and understood by the lay person. To assist in this, most chapters begin with a simplified "Synopsis". There may be portions of this work that will take some effort to thoroughly understand the principles addressed. If you find yourself in one of those places, continue to read and gain what understanding you can. A second visit to those sections, will undoubtedly increase the depth of your understanding.

The first book in this series entitled: "Fatal Flaws; How the Assumptions, Materials and Methods Invalidate the Theory of Global Warming" has been published by Gatekeeper Press and is available in paperback and eBook formats from their associated distributors (Kindle, Apple Books, Barnes & Nobel and other book sellers).

For more on these topics visit: <fatalflaws.net> You may reach the author at: earljseeley@gmail.com

www.ingramcontent.com/pod-product-compliance
Ingram Content Group UK Ltd.
Pitfield, Milton Keynes, MK11 3LW, UK
UKHW062256290726
14090UKWH00017B/733